Lecture Notes in Computer Science **14203**

The series Lecture Notes in Computer Science (LNCS), including its subseries Lecture Notes in Artificial Intelligence (LNAI) and Lecture Notes in Bioinformatics (LNBI), has established itself as a medium for the publication of new developments in computer science and information technology research, teaching, and education.

LNCS enjoys close cooperation with the computer science R & D community, the series counts many renowned academics among its volume editors and paper authors, and collaborates with prestigious societies. Its mission is to serve this international community by providing an invaluable service, mainly focused on the publication of conference and workshop proceedings and postproceedings. LNCS commenced publication in 1973.

Shunli Zhang · Bo Hu · Liang-Jie Zhang
Editors

Big Data – BigData 2023

12th International Conference
Held as Part of the Services Conference Federation, SCF 2023
Honolulu, HI, USA, September 23–26, 2023
Proceedings

Editors
Shunli Zhang
Jiujiang University
Jiangxi, China

Bo Hu
Shenzhen Yihuo Technology Co., Ltd.
Shenzhen, China

Liang-Jie Zhang
Shenzhen Entrepreneurship and Innovation Federation
Shenzhen, China

ISSN 0302-9743 ISSN 1611-3349 (electronic)
Lecture Notes in Computer Science
ISBN 978-3-031-44724-2 ISBN 978-3-031-44725-9 (eBook)
https://doi.org/10.1007/978-3-031-44725-9

This Springer imprint is published by the registered company Springer Nature Switzerland AG
The registered company address is: Gewerbestrasse 11, 6330 Cham, Switzerland

Preface

The 2023 International Congress on Big Data (BigData 2023) aimed to provide an international forum to formally explore various business insights into all kinds of value-added "services". Big Data is a key enabler in exploring business insights and economics of services.

BigData 2023 was a member of the Services Conference Federation (SCF). SCF 2023 had the following 10 collocated service-oriented sister conferences: 2023 International Conference on Web Services (ICWS 2023), 2023 International Conference on Cloud Computing (CLOUD 2023), 2023 International Conference on Services Computing (SCC 2023), 2023 International Conference on Big Data (BigData 2023), 2023 International Conference on AI & Mobile Services (AIMS 2023), 2023 International Conference on Metaverse (Metaverse 2023), 2023 International Conference on Internet of Things (ICIOT 2023), 2023 International Conference on Cognitive Computing (ICCC 2023), 2023 International Conference on Edge Computing (EDGE 2023), and 2023 International Conference on Blockchain (ICBC 2023).

This volume presents the accepted papers for BigData 2023, held in Hawaii and as a virtual conference, during September 23–26 2023. BigData 2023's major topics included but were not limited to: Big Data Architecture, Big Data Modeling, Big Data as a Service, Big Data for Vertical Industries (Government, Healthcare, etc.), Big Data Analytics, Big Data Toolkits, Big Data Open Platforms, Economic Analysis, Big Data for Enterprise Transformation, Big Data in Business Performance Management, Big Data for Business Model Innovations and Analytics, Big Data in Enterprise Management Models and Practices, Big Data in Government Management Models and Practices, and Big Data in Smart Planet Solutions.

We received 27 submissions, and accepted 16 papers. Each was reviewed and selected by at least three independent members of the BigData 2023 International Program Committee in a single-blind review process. We are pleased to thank the authors whose submissions and participation made this conference possible. We also want to express our thanks to the Organizing Committee and Program Committee members, for their dedication in helping to organize the conference and reviewing the submissions. We appreciate your great contributions as volunteers, authors, and conference participants in the fast-growing worldwide services innovations community.

September 2023

Shunli Zhang
Bo Hu
Liang-Jie Zhang

Preface

Organization

General Chair

Bo Hu	Shenzhen Yihuo Technology Co., Ltd., China

Program Chair

Shunli Zhang	Jiujiang University, China

Services Conference Federation (SCF 2023)

General Chairs

Ali Arsanjani	Google, USA
Wu Chou	Essenlix Corporation, USA

Coordinating Program Chair

Liang-Jie Zhang	Shenzhen Entrepreneurship & Innovation Federation, China

CFO and International Affairs Chair

Min Luo	Georgia Tech, USA

Operation Committee

Jing Zeng	China Gridcom Co., Ltd., China
Yishuang Ning	Tsinghua University, China
Sheng He	Tsinghua University, China

Steering Committee

Calton Pu (Co-chair)	Georgia Tech, USA
Liang-Jie Zhang (Co-chair)	Shenzhen Entrepreneurship & Innovation Federation, China

BigData 2023 Program Committee

Peter Baumann	Jacobs University, Germany
Wei Dong	Ann Arbor Algorithms Inc., USA
Yu Liang	University of Tennessee at Chattanooga, USA
Anand Nayyar	Duy Tan University, Vietnam
Kai Peng	Huaqiao University, China
Wenbo Wang	GoDaddy, USA
Daqing Yun	Harrisburg University, USA
Manas Gaur	Wright State University, USA
Sagar Sharma	Wright State University, USA
Nan Wang	Heilongjiang University, China
Jinzhu Gao	University of the Pacific, USA
Harald Kornmayer	DHBW Mannheim, Germany
Luiz Angelo Steffenel	Université de Reims Champagne-Ardenne, France
Jing Zeng	China Gridcom Co., Ltd., China

Conference Sponsor – Services Society

The Services Society (S2) is a non-profit professional organization that has been created to promote worldwide research and technical collaboration in services innovations among academia and industrial professionals. Its members are volunteers from industry and academia with common interests. S2 is registered in the USA as a "501(c) organization", which means that it is an American tax-exempt nonprofit organization. S2 collaborates with other professional organizations to sponsor or co-sponsor conferences and to promote an effective services curriculum in colleges and universities. S2 initiates and promotes a "Services University" program worldwide to bridge the gap between industrial needs and university instruction.

The Services Sector accounted for 79.5% of the GDP of the USA in 2016. The Services Society has formed 5 Special Interest Groups (SIGs) to support technology and domain specific professional activities.

- Special Interest Group on Services Computing (SIG-SC)
- Special Interest Group on Big Data (SIG-BD)
- Special Interest Group on Cloud Computing (SIG-CLOUD)
- Special Interest Group on Artificial Intelligence (SIG-AI)
- Special Interest Group on Metaverse (SIG-Metaverse)

About the Services Conference Federation (SCF)

As the founding member of the Services Conference Federation (SCF), the first **International Conference on Web Services (ICWS)** was held in June 2003 in Las Vegas, USA. Meanwhile, the First International Conference on Web Services - Europe 2003 (ICWS-Europe 2003) was held in Germany in October 2003. ICWS-Europe 2003 was an extended event of the 2003 International Conference on Web Services (ICWS 2003) in Europe. In 2004, ICWS-Europe was changed to the European Conference on Web Services (ECOWS), which was held in Erfurt, Germany. Sponsored by the Services Society and Springer, SCF 2018 and SCF 2019 were held successfully in Seattle and San Diego, USA. SCF 2020 and SCF 2021 were held successfully online and in Shenzhen, China. SCF 2022 was held successfully in Hawaii, USA. To celebrate its 21st birthday, SCF 2023 was held on September 23–26, 2023, in Honolulu, Hawaii, USA with Satellite Sessions in Shenzhen, Guangdong, China.

In the past 20 years, the ICWS community has been expanded from Web engineering innovations to scientific research for the whole services industry. The service delivery platforms have been expanded to mobile platforms, Internet of Things, cloud computing, and edge computing. The services ecosystem has gradually been enabled, value added, and intelligence embedded through enabling technologies such as big data, artificial intelligence, and cognitive computing. In the coming years, all transactions with multiple parties involved will be transformed to blockchain.

Based on technology trends and best practices in the field, the Services Conference Federation (SCF) will continue serving as the conference umbrella's code name for all services-related conferences. SCF 2023 defined the future of New ABCDE (AI, Blockchain, Cloud, BigData & IOT) and enters the 5G for Services Era. The theme of SCF 2023 was **Metaverse Era**. We are very proud to announce that SCF 2023's 10 co-located theme topic conferences all centered around "services", while each focused on exploring different themes (web-based services, cloud-based services, Big Data-based services, services innovation lifecycle, AI-driven ubiquitous services, blockchain-driven trust service-ecosystems, industry-specific services and applications, and emerging service-oriented technologies).

- Bigger Platform: The 10 collocated conferences (SCF 2023) were sponsored by the Services Society, which is the world-leading not-for-profit organization (501 c(3)) dedicated to the service of more than 30,000 worldwide Services Computing researchers and practitioners. A bigger platform means bigger opportunities for all volunteers, authors, and participants. Meanwhile, Springer provided sponsorship to best paper awards and other professional activities. All the 10 conference proceedings of SCF 2023 will be published by Springer and indexed in ISI Conference Proceedings Citation Index (included in Web of Science), Engineering Index EI (Compendex and Inspec databases), DBLP, Google Scholar, IO-Port, MathSciNet, Scopus, and ZBlMath.

- Brighter Future: While celebrating the 2023 version of ICWS, SCF 2023 highlighted the Second International Conference on Metaverse (METAVERSE 2023), which covered immersive services for all vertical industries and area solutions. Its focus was on industry-specific services for digital transformation. This will lead our community members to create their own brighter future.
- Better Model: SCF 2023 will continue to leverage the invented Conference Blockchain Model (CBM) to innovate the organizing practices for all the 10 theme conferences. Senior researchers in the field are welcome to submit proposals to serve as CBM Ambassador for an individual conference to start better interactions during your leadership role in organizing future SCF conferences.

Contents

Research Track

Application Track

Short Paper Track

Research Track

A Survey on Cross-Domain Few-Shot Image Classification

Shisheng Deng[1,2], Dongping Liao[3], Xitong Gao[1](✉), Juanjuan Zhao[1], and Kejiang Ye[1]

[1] Shenzhen Institute of Advanced Technology, Chinese Academy of Sciences, Shenzhen 518000, China
{ss.deng, xt.gao, jj.zhao,kj.ye}@siat.ac.cn
[2] University of Chinese Academy of Sciences, Beijing 100049, China
[3] University of Macau, Macau SAR 999078, China
yb97428@um.edu.mo

Abstract. Due to the limited availability of labelled data in many real-world scenarios, we have to resort to data from other domains to improve models' performance, which prompts the advancement of research regarding the cross-domain few-shot image classification task. In this paper, we systematically review existing cross-domain few-shot image classification algorithms published in recent years. We categorize these algorithms into data-augmentation and feature-alignment paradigms and present their recent progress. We summarize three commonly-used cross-domain datasets for benchmarking few-shot image classification tasks and relevant scenarios. Finally, we outline existing limitations and future perspectives.

Keywords: Deep Learning · Few-shot Image Classification · Cross-Domain

1 Introduction

Over the last decade, deep learning (DL) [30] has achieved excellent results on many application scenarios, including computer vision [20], natural language processing [14], *etc.* Traditional DL methods are not effective in tasks with limited training data. In contrast, humans can leverage their accumulated knowledge to quickly learn the characteristics of unfamiliar things with a limited amount of data. To address this issue, researchers have introduced the concept of Few-Shot Learning (FSL) [57]. FSL aims to mimic the human learning process and achieve better generalization performance by using a limited number of training samples in scenarios where data is scarce. Recently, Few-Shot Image Classification (FSIC) [57] algorithms have demonstrated better classification accuracy than humans in image classification. However, these remarkable outcomes are limited

S. Deng and D. Liao—Equal contribution.

S. Zhang et al. (Eds.): BigData 2023, LNCS 14203, pp. 3–17, 2023.
https://doi.org/10.1007/978-3-031-44725-9_1

to scenarios where there is only a slight difference between the distribution of the training data and the test data. For situations where there is a sizeable distributional difference between the training and test data, the model will suffer significant performance degradation due to the discrepancy between the different domains. Researchers have thus formalized the Cross-Domain Few-Shot Image Classification (CDFSIC) [7], along with its corresponding classification algorithms to investigate the challenges in cross-domain few-shot learning.

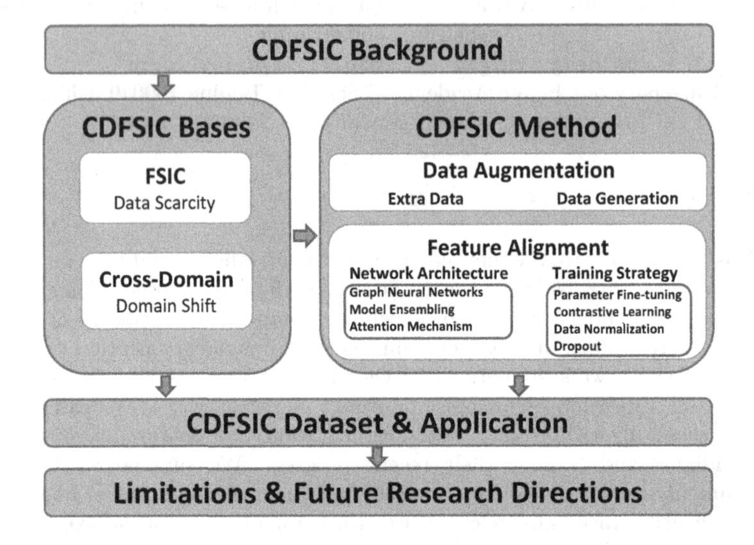

Fig. 1. The framework of survey.

This paper presents a thorough and systematic review of CDFSIC. As shown in Fig. 1, the survey is structured as follows. First, following the introduction of CDFSIC in this section, we present the preliminaries of CDFSIC in Sect. 2, which includes the definitions of FSIC and Cross-Domain problems. We then provide a summary of the current CDFSIC methods, including an introduction to standard datasets and applications. Finally, we discuss the limitations and challenges of CDFSIC that may present future research opportunities.

2 Preliminaires of CDFSIC

2.1 Few-Shot Image Classification

Few-Shot Learning (FSL) [57] is a machine learning technique that involves training a model to achieve strong generalization performance using only a limited number of training examples. One of the most widely-used benchmark for evaluating FSL algorithms is Few-Shot Image Classification (FSIC), which has numerous realistic applications [57].

A FSIC task can be defined as $\mathcal{D}_{\text{FSIC}} = \{\mathcal{D}_{\text{train}}, \mathcal{D}_{\text{test}}\}$, where$\{y \mid (x, y) \in \mathcal{D}_{\text{train}}\} \cap \{y \mid (x, y) \in \mathcal{D}_{\text{test}}\} = \emptyset$, *i.e.*, the test and train datasets do not contain common labels. Following [29], most recent works on FSIC employ the standard *N-way K-shot* (*M-query*) episodic task learning.

Specifically, for each FSIC task, we sample n episodic tasks $\{T_1, \ldots, T_n\}$ from $\mathcal{D}_{\text{train}}$ as training episodes, and m episodic tasks $\{T_1, \ldots, T_m\}$ from $\mathcal{D}_{\text{test}}$ as testing episodes. Each episodic task T_i consists of a support set T_i^S and a query set T_i^Q. From a dataset, each episodic tasks randomly samples N categories respectively, with each category sampling K image-label pairs (x, y), $T_i^S = \{(x_k, y_k)\}_{k=1}^{N \times K}$ for support set, and each category sampling M image-label pairs (x, y), $T_i^Q = \{(x_k, y_k)\}_{k=1}^{N \times M}$ for query set. Both $\mathcal{D}_{\text{train}}$ and $\mathcal{D}_{\text{test}}$ samples the support and query sets following the above configuration, except that the $\mathcal{D}_{\text{test}}$ provides no labels for the query set, namely, $T_i^Q = \{(x_k)\}_{k=1}^{N \times M}$.

2.2 The Cross-Domain Problem

Blanchard *et al.* [3] formally presented the Cross-Domain (CD) problem in machine learning, while Torralba *et al.* [47] brought research attention to the cross-domain problem in computer vision tasks. They investigated the performance of classification models by thorough evaluation on six popular benchmark datasets. Their experiments showed that the intrinsic dataset bias introduced by the domain gap will lead to poor generalization performance.

A domain is defined as a joint distribution $P(X, Y)$ [70] of the input (data) space X and output (label) space Y. For the Cross-Domain problem, the source-domain distribution $P_S(X, Y)$ and the target-domain distribution $P_T(X, Y)$ are notably different. Moreover, the data of target domain is not available during the model training process. Most of the research has focused on the multi-source scenario, which presupposes the availability of several distinct yet relevant domains. Specifically, given K similar but distinct source domains, $S = \{S_k = \{(x^k, y^k)\}\}_{k=1}^{K}$, each domain is represented by a joint distribution $P_S^k(X, Y)$. Note that $P_S^k(X, Y)$ is dissimilar to $P_S^{k'}(X, Y)$, with $k \neq k'$ for $k, k' \in \{1, \cdots, K\}$. The joint distribution corresponding to the target domain is denoted as $P_T(X, Y)$. In addition, $P_T(X, Y)$ is also dissimilar to $P_S^k(X, Y)$, where $k \in \{1, \cdots, K\}$.

The *cross-domain few-shot image classification* (CDFSIC) problem, first introduced by Chen *et al.* [7], poses challenges of both Cross-Domain and Few-Shot Image Classification, including a scarce sample size and considerable differences between the training and testing data distributions. The models trained under CDFSIC would thus require stronger generalization capabilities than traditional FSIC models for better adaptation to novel target domains.

3 CDFSIC Algorithm

In general, CDFSIC faces two challenges: data scarcity and domain shift. Based on these challenges, the current approach of CDFSIC can be categorized into two camps: *data augmentation* and *feature alignment* methods.

3.1 Data Augmentation Methods

Data augmentation [45], commonly utilized in deep learning methods, can mitigate the possibility of overfitting, which may happen when the training dataset has a limited number of samples, while having low diversity. Recently, some researchers employ additional larger datasets (*e.g.*, ImageNet [12]) as training data to augment the FSIC task. This technique aims to learn valuable features from a varied dataset with higher diversity [18]. Additionally, data generation [45] is another popular data augmentation technique. Based on these approaches, we categorize data augmentation methods into two: *extra data* and *data generation.*

Extra Data. As part of their work, Chen *et al.* [7] introduced the first benchmark dataset for the CDFSIC task, namely MiniImageNet → CUB. They employed MiniImagenet [52] as the source domain, which is relatively similar to the target domain, CUB [53].

Real-world CDFSIC scenarios involve domains that differ greatly in data volume and distribution. Addressing this issue, Guo *et al.* [18] proposed a broader CDFSIC baseline than previous work. Employing ImageNet as the source domain, they conduct experiments on four datasets with varying degrees of similarity to the natural image based on 3 orthogonal criteria: 1) existence of perspective distortion, 2) the semantic content, and 3) color depth. Experiments showed that the accuracy of CDFSIC methods is dependent on the degree of similarity between the source and target domain. While Chen *et al.* [7] proposed a 2-stage training approach (pretrain → metatrain), Hu *et al.* [24] introduced a 3-stage training pipeline (pretrain → metatrain → finetune). Hu *et al.* also evaluated the effectiveness of various feature extraction networks and showed that Vision Transformer [27] performs better than standard convolutional networks [37] and residual networks (ResNets) [20].

Compared to traditional FSIC approaches, methods that leverage extra data are useful but computationally demanding. Therefore, data generation methods that are less computationally intensive have been introduced for the CDFSIC task.

Data Generation. Data generation refers to generating new labeled data through commonly-used data synthesis techniques, such as MixUp [63], geometric transformations [45], *etc.*

Fu *et al.* [16] propose a feature-wise domain adaptation module called Feature Distribution Matching (FDM) to guide the MixUp process. FDM measures the discrepancy between the feature distributions of the source and target domain and encourages the model to generate synthetic samples that are more similar to the target domain. Zhang *et al.* [64] and Deng *et al.* [13] apply rotation transformations to images and predict the rotation angle in the pretrain phase. Mazumder *et al.* [34] proposed the composite rotation auxiliary task as a data generation method for the CDFSIC task. This method involves two levels of rotation on the image: first, rotating patches within the image (inner rotation); and then rotating the entire image (outer rotation) before assigning a rotation class to the transformed image for the model to learn to predict via self-supervision.

Although data generation methods require less computing effort and are easy to implement, they face limitations in significantly improving classification accuracy since the generated samples are derived from the original dataset. Therefore, while data generation methods may be used to boost accuracies of the CDFSIC task, their performance is relatively limited when compared to methods that utilize additional training data.

3.2 Feature Alignment Methods

To address data scarcity issue in CDFSIC, data augmentation based method essentially enhances the diversity of samples by expanding the sample space. To handle the problem of domain shift [56] in CDFSIC, feature alignment methods aims to align the features extracted from the source domain with those extracted from the target domain. We summarize the existing feature alignment based method by casting them into two categories: *network architecture design* and *training strategy improvement.*

Network Architecture Design. Network architecture design refers to designing or refining the model structure to enhance the ability of the model to generalize the source domain feature characteristics to the target domain. We summarize the existing network architecture design methods as follows:

- **Graph Neural Networks** (GNN) [44] are widely used in graph analysis due to their better scalability and interpretation comparing to traditional graph learning algorithms, such as, Graph Signal Processing, Random Walk and Matrix Factorization. In FSIC, researchers usually take an image as a node of the GNN, while the similarity of image pairs is considered as an edge of the GNN [43]. GNN-based methods parameterize the metric function in FSIC task, allowing a closer fit to the realistic metric function between image pairs. A number of excellent works have emerged in traditional FSIC tasks [28,43,59], and CDFSIC.
  To alleviate the issue of information loss with the increasing number of the GNN layer and improve the graph-structured data features representation quality, Liu *et al.* [33] propose a geometric algebra graph neural network (GA-GNN) that maps graph nodes to a high-dimensional geometric algebraic space, allowing for a better measurement of the discrepancy between image pairs. Chen *et al.* [8] introduce a Flexible Graph Neural Network (FGNN) that adaptively selects the node feature dimensions to enhance the relevance between image pairs. Most current methods for domain alignment focus on utilizing local spatial information while neglecting the strong correspondence of non-local spatial information (non-local relationships). Accordingly, Zhang *et al.* [67] present a Dual Graph Cross-domain Few-shot Learning (DG-CFSL) framework to learn the domain distribution properties and mitigate

the domain shift, specifically, optimize the dual graph, feature graph and distribution graph simultaneously to achieve domain alignment.
The fundamental concept of the CDFSIC methods based on GNNs is to iteratively update the node features and deduce the relationships between nodes. It features strong interpretability [43] and exhibits great classification performance, but demands significant computational and memory resources. As every two images require the construction of an edge, the memory and computational cost will increase quadratically with the number of samples during inference. Therefore, in CDFSIC tasks, GNN-based method still suffer from the aforementioned limitations that merits further research and improvement.
- **Model Ensembling** [42] is considered as the state-of-the-art solution for many machine learning challenges, aiming to merge multiple models in some way (*e.g.*, voting, averaging, stacking, *etc.*) to extract their strengths and improve the generalization performance of the final model.
Liu *et al.* [31] have put forth a proposal for the CDFSIC task, which involves using an ensemble model with feature transformation. Specifically, they suggested constructing a prediction model by performing diverse feature transformations after extracting features using a network. While Liu *et al.* [31] ensemble the feature extractor, Adler *et al.* [1] integrate from the classifier perspective. In CDFSIC, domain shifts can cause a significant divergence in high-level concepts between the source and target domain. However, low-level concepts, such as image edges, may still retain relevance and applicability. To tackle the challenge, Adler *et al.* [1] introduce a novel approach called Cross-domain Hebbian Ensemble Few-shot learning (CHEF) that utilizes an ensemble of Hebbian learners, which operate on different layers of a deep neural network to merge representations. Through the fusion process, CHEF facilitates the transfer of useful low-level features while accommodating high-level concept shifts.
In CDFSIC tasks, ensemble of multiple models trained across different scenarios can equip algorithms with diverse knowledge of various scenes, effectively addressing the issue of limited generalization ability of models. However, it is important to note that the training of ensembles incurs significant computational and storage costs that increase linearly with the number of scenarios.
- The **Attention Mechanism** [5] in neural networks draws inspiration from the physiological perception of the environment by humans. For example, our visual system tends to selectively focus on certain parts of the visual field while disregarding irrelevant information. Similarly, in various natural language scenarios, some parts of the input to the model are more important than others. The attention mechanism allows for the selective processing of model features, enhancing the model's generalization performance.
Hou *et al.* [22] propose a novel attention module to tackle the problem of generalization to novel classes, known as the Cross Attention Module (CAM). The CAM generates cross attention maps for each pair of class feature and query sample feature, with the aim of highlighting the relevant object regions and enhancing the discriminative power of the extracted features. The innovative method shows promising results in improving the performance of various

computer vision tasks, particularly in scenarios where generalization to new categories is required. Ye *et al.* [62] introduce an innovative attention method to customize instance embeddings for a given classification task using a set-to-set function. This approach generates task-specific embeddings that are also highly discriminative. To determine the most effective set-to-set functions, they conducted empirical investigations on several variations and discovered that the Transformer [27] was the best option. This is because the Transformer inherently satisfies the key properties required for the desired model. According to Liu *et al.* [32], model ensemble is an effective method for tackling the CDFSIC task. However, when combining models trained on different domains, it is important to take into account that the ratio of model parameter weights should not be equal in the final model. To address this issue, they propose a task-adaptive model weight method, which involves fixing the parameters of all feature extractors after training on the source domain, and subsequently training an attention structure. Sa *et al.* [41] present a simple and effective model for Attentive Fine-Grained Recognition (AFGR). They introduce a residual attention module (RAM) [54] that is integrated into the feature encoder of the residual network. This module enhances various semantic features linearly, enabling the metric function to locate fine-grained feature information better in an image.
Attention mechanism has been demonstrated effective to enhance the interpretability of CDFSIC algorithms and improve the semantic representation capabilities of models. As such, we believe that there is still considerable untapped potential for its application in this field. One potential future research direction is to explore the combination of attention mechanism with feature disentanglement [40] to propose more sophisticated and effective attention mechanisms. By doing so, we can further improve the accuracy and interpretability of CDFSIC methods.

Training Strategy Improvement. Training strategy improvement refers to improving the model performance during the model training process to align the source domain features with the target domain features. We summarize the existing training strategies as follows:

- **Parameter Fine-tuning** [23] is a machine learning technique that involves modifying the parameters of a pretrained model to adapt it to a new dataset while focusing on a specific task.
  Chen *et al.* [7] propose two simple baselines, which provides the first evidence of the powerful capabilities of fine-tuning in CDFSIC. Similarly, Guo *et al.* [18] use a straightforward fine-tuning approach but differed from Chen *et al.* [7] by fixing the low-dimensional feature layer of the feature extractor during fine-tuning on the target domain on the last three layers. Meanwhile, Cai *et al.* [4] propose a meta fine-tuning mechanism, which utilizes a meta-learning [15] approach to initialize the weights that need to be fine-tuned, rather than directly fine-tuning an incompletely pretrained model. Reinitialization [65] has been widely explored in the natural language field, especially

in the BERT [14] model. Oh *et al.* [35] propose a method for CDFSIC that involves re-initializing the final residual block of the feature extractor before fine-tuning on the target domain. This is done after supervised training on the source domain. This approach reduces learning bias towards the source domain by simply re-initializing specific layers for a given domain, providing a fresh perspective for fine-tuning on CDFSIC.
Fine tuning the parameters of a model can rapidly assist it in adapting to new scenarios and effectively align the features of both the source and target domains, making it a crucial technique for tackling cross-domain issues. In the case of CDFSIC tasks, there is still ample scope for further research in parameter fine-tuning.

- **Contrastive Learning.** In recent years, a new paradigm of Self-Supervised Learning (SSL) [26] called Contrastive Learning (CL) [36] has emerged as an effective tool for unsupervised learning. CL generates a similarity distribution of data by comparing pairs of samples, and adjusts the model parameters accordingly. By optimizing the contrastive loss [19], the model is encouraged to extract more similar features from pairs of samples in the same class, while features from pairs of samples in different classes are encouraged to be more disperse.
Zhang *et al.* [66] employ the AmdimNet [6] as backbone for training, which utilizes contrastive loss maximization on the mutual information between two new views generated from the same image. Das *et al.* [10] propose a Contrastive Learning and Feature Selection System (ConFeSS) for CDFSIC. ConFeSS optimizes in pretrain stage by contrastive loss and fine-tunes using sample pairs with masked relevant classification features to addresses the issue of overfitting and achieves improved performance. In order to mitigate overfitting, Das *et al.* [11] propose a new fine-tuning method that relies on contrastive loss. This approach utilizes unlabelled examples from the source domain as distractors, which serves to repurpose them and prevent overfitting.
In the CDFSIC, the use of contrastive loss can enhance model's ability to generalize by effectively leveraging the representation in unlabelled data to pull together intra-class samples and push apart inter-class ones. As a result, contrastive loss holds practical value in realistic scenarios where ample unlabelled data is available. However, due to the absence of explicit supervision, contrastive loss is susceptible to problems such as slow convergence and instability, necessitating further investigation.
- **Data Normalization** [46] is a crucial technique in data processing that involves mapping data into a common scale. It is especially important when dealing with data from different sources, as it allows for easier comparison and analysis. In the context of CDFSIC, images from the source and target domains usually exhibit significant differences in terms of style, color, and quality. These differences could have a negative impact on the model's ability to generalize well to new data.
Wang *et al.* [55] and Xu *et al.* [58] both normalize the extracted image features before classification to reduce the discrepancy between samples from the source and target domains. However, they employ different normaliza-

tion techniques. Wang *et al.* [55] standardize the feature vectors using 1, 2, 3, and ∞ p-norms, while Xu et al. [58] use two learnable parameters γ, β for Instance Normalization $IN(F) = \gamma\frac{F-\mu(F)}{\sigma(F)} + \beta$, where F refer to the image feature, $\mu(\cdot)$ and $\sigma(\cdot)$ denote the mean and standard deviation calculated at the channel level for each sample. Yazdanpanah *et al.* [60,61] and Tseng *et al.* [49] make improvements to the Batch Normalization (BN) Layer in the feature extraction network. According to Yazdanpanah *et al.* [61], the use of trainable parameters in the BN layer of convolutional neural networks will lead to a shift in the distribution of batch data, while also improving the convergence rate during training on the source domain. However, it may not generalize well to the target domain, which can limit classification performance. To address the issue, Yazdanpanah *et al.* [61] replaced the BN layer in the convolutional network with a Feature Normalization (FN) layer, $FN(h_c) = \frac{h_c-\mu_c}{\sqrt{\sigma_c^2+\epsilon}}$, Here, h_c denotes batch data feature, μ_c and σ_c are the first and second moments [38] of h_c. In contrast to the BN layer, the FN layer discards the trainable parameters for shifting and scaling. In their subsequent work, Yazdanpanah *et al.* [60] propose that the parameters within the BN layer are trained using source domain data, leading to a potential mismatch between the internal BN parameters and the data distribution during inference caused by domain shift. To tackle the issue, they introduce a Visual Domain Bridge (VDB) that replaces the statistical mean and variance of the target domain data with those of the source domain, generating a transformed data feature, then fine-tune the model using the transformed feature to alleviate the mismatch between the BN layer's internal parameters and the target domain's data distribution. Tseng *et al.* [49] propose adding a Feature-Wise Transformation (FWT) layer after the BN layer in convolutional neural networks to simulate feature distributions in different domains, improving the generalization ability of the feature extractor.
Data normalization is crucial for improving image classification accuracy. It helps the model converge in cross-domain scenarios and aligns the feature distributions of the source and target domains by reducing distribution discrepancies. Therefore, data normalization is a practical method to enhance the generalization ability of the model in CDFSIC task.

- **Dropout** is a commonly-used technique in deep learning to regularize training. Hinton *et al.* [21] point out that over-parameterization of the model can easily lead to overfitting, while dropout can effectively alleviate overfitting and to some extent act as regularization, improving the performance of the network.
  According to Huang *et al.* [25], dropout can be a useful technique in CDFSIC. By dropping out the activations of the most important features in the training data, the network is forced to activate the second most important features that are related to the labels. This approach can effectively unlock the potential of the network, leading to enhanced generalization performance. Tu *et al.* [50] propose a simple and effective dropout-style method to enhance model trained on low-complexity concepts from the source domain. The app-

roach involves sampling multiple sub-networks by dropping neurons or feature maps to create a diverse set of models with varied features for the target domain. The most suitable sub-networks are selected to form an ensemble for target domain learning. This method enables the model to generalize better to the target domain, where it may encounter novel and complex concepts. In conclusion, dropout can effectively alleviate overfitting on CDFSIC task without increasing computational or memory overhead.

4 CDFSIC Dataset and Application

4.1 Standard Datasets

Currently, in CDFSIC, the datasets used in different literature are not entirely consistent. Table 1 shows three commonly-used benchmark datasets.

Table 1. Standard Dataset of CDFSIC

Dataset	Published In	Code/Data Link
MiniImageNet → CUB [7]	ICLR 19	https://github.com/wyharveychen/CloserLookFewShot
BSCDFSL [18]	ECCV 20	https://github.com/IBM/cdfsl-benchmark
Meta-Dataset [48]	ICLR 20	https://github.com/google-research/meta-dataset

MiniImageNet → CUB and BSCDFSL are widely-used datasets in recent works. Due to the late release of MetaDataset, there are only a few works evaluated on this dataset.

4.2 CDFSIC Application

CDFSIC algorithms have already found applications in various fields, including medical imaging such as X-ray images [9], skin disease images [17], and satellite remote sensing images [2] as well as hyperspectral images [68]. Moreover, we foresee that CDFSIC algorithms have immense potential in other domains, such as aerospace, cultural heritage preservation, and public safety.

5 Limitations and Future Research Directions

In recent years, there are some advancements in addressing the problem of CDFSIC, particularly on challenges related to data scarcity and domain shift between source and target domain. However, despite these developments, there are still other limitations that need to be overcome in this field.

5.1 Limitations of the Current FSIC Settings

Currently, FSIC tasks generally follow N-*way* K-*shot* (M-*query*) setting, where N refers to the number of image categories in a sub-task, and K refers to the number of samples in each category contained in the support set. N-*way* K-*shot* setting is reasonable for real-world scenarios because the number of samples for each category in the support set can be artificially set when creating the dataset. However, in testing phase, the number of samples for each category in the query set may not be the same, denoted by M. Furthermore, we cannot predict the distribution of the query data easily, nor can we assume that it is evenly distributed among each category.

Veilleux *et al.* [51] propose to use Dirichlet Distribution to simulate imbalanced sample distribution for each category in the query set of a sub-task, making it closer to real-world scenarios. We believe that addressing imbalanced FSIC is an important area of future research.

5.2 Theoretical Insights

In the field of CDFSIC, current state-of-the-art algorithms are usually developed through empirical exploration, without sufficient theoretical guidance. For traditional FSIC tasks, various theoretical derivations have been proposed [15,39]. However, for CDFSIC, current research merely combines traditional FSIC naively with cross-domain techniques. Therefore, there is an urgent need for future research that provides theoretical support for CDFSIC.

5.3 Cross-Hardware CDFSIC

In addition to the CDFSIC issues mentioned above, Zhao *et al.* [69] further explore the cross-hardware scenario of FSIC, optimizing the inference latency of the model on hardware devices such as GPUs, ASICs, and IoT platforms. As cross-domain scenarios do not require training and testing data to have consistent distributions, we anticipate that it is even more necessary for CDFSIC algorithms to optimize performance for hardware in order to meet its wider application prospects.

6 Conclusion

In the field of image classification, research on FSIC has recently extended to CDFSIC. This paper provides a detailed overview of the current state of research on CDFSIC, while analyzing the challenges faced by such research and providing a perspective on its future prospects.

Acknowledgments. This work is supported in part by National Key R&D Program of China (No. 2019YFB2102100), Key-Area Research and Development Program of Guangdong Province (No. 2020B010164003), and Shenzhen Science and Technology Innovation Commission (No. JCYJ20190812160003719).

References

1. Adler, T., et al.: Cross-domain few-shot learning by representation fusion. arXiv preprint arXiv:2010.06498 (2020)
2. Ammour, N., Bashmal, L., Bazi, Y., Al Rahhal, M.M., Zuair, M.: Asymmetric adaptation of deep features for cross-domain classification in remote sensing imagery. IEEE Geosci. Remote Sens. Lett. **15**(4), 597–601 (2018)
3. Blanchard, G., Lee, G., Scott, C.: Generalizing from several related classification tasks to a new unlabeled sample. In: Advances in Neural Information Processing Systems, vol. 24 (2011)
4. Cai, J., Shen, S.M.: Cross-domain few-shot learning with meta fine-tuning. arXiv preprint arXiv:2005.10544 (2020)
5. Chaudhari, S., Mithal, V., Polatkan, G., Ramanath, R.: An attentive survey of attention models. ACM Trans. Intell. Syst. Technol. (TIST) **12**(5), 1–32 (2021)
6. Chen, D., Chen, Y., Li, Y., Mao, F., He, Y., Xue, H.: Self-supervised learning for few-shot image classification. In: ICASSP 2021–2021 IEEE International Conference on Acoustics, Speech and Signal Processing (ICASSP), pp. 1745–1749. IEEE (2021)
7. Chen, W.Y., Liu, Y.C., Kira, Z., Wang, Y.C.F., Huang, J.B.: A closer look at few-shot classification. In: International Conference on Learning Representations (2018)
8. Chen, Y., et al.: Cross-domain few-shot classification based on lightweight res2net and flexible GNN. Knowl.-Based Syst. **247**, 108623 (2022)
9. Cohen, J.P., Hashir, M., Brooks, R., Bertrand, H.: On the limits of cross-domain generalization in automated x-ray prediction. In: Medical Imaging with Deep Learning, pp. 136–155. PMLR (2020)
10. Das, D., Yun, S., Porikli, F.: Confess: a framework for single source cross-domain few-shot learning. In: International Conference on Learning Representations (2022)
11. Das, R., Wang, Y.X., Moura, J.M.: On the importance of distractors for few-shot classification. In: Proceedings of the IEEE/CVF International Conference on Computer Vision, pp. 9030–9040 (2021)
12. Deng, J., Dong, W., Socher, R., Li, L.J., Li, K., Fei-Fei, L.: ImageNet: a large-scale hierarchical image database. In: 2009 IEEE Conference on Computer Vision and Pattern Recognition, pp. 248–255. IEEE (2009)
13. Deng, S., Liao, D., Gao, X., Zhao, J., Ye, K.: Improving few-shot image classification with self-supervised learning. In: Ye, K., Zhang, L.J. (eds.) CLOUD 2022. LNCS, vol. 13731, pp. 54–68. Springer, Cham (2022). https://doi.org/10.1007/978-3-031-23498-9_5
14. Devlin, J., Chang, M.W., Lee, K., Toutanova, K.: BERT: pre-training of deep bidirectional transformers for language understanding. arXiv preprint arXiv:1810.04805 (2018)
15. Finn, C., Abbeel, P., Levine, S.: Model-agnostic meta-learning for fast adaptation of deep networks. In: International Conference on Machine Learning, pp. 1126–1135. PMLR (2017)
16. Fu, Y., Fu, Y., Jiang, Y.G.: Meta-fdmixup: cross-domain few-shot learning guided by labeled target data. In: Proceedings of the 29th ACM International Conference on Multimedia, pp. 5326–5334 (2021)
17. Gu, Y., Ge, Z., Bonnington, C.P., Zhou, J.: Progressive transfer learning and adversarial domain adaptation for cross-domain skin disease classification. IEEE J. Biomed. Health Inform. **24**(5), 1379–1393 (2019)

18. Guo, Y., et al.: A broader study of cross-domain few-shot learning. In: Vedaldi, A., Bischof, H., Brox, T., Frahm, J.-M. (eds.) ECCV 2020, Part XXVII. LNCS, vol. 12372, pp. 124–141. Springer, Cham (2020). https://doi.org/10.1007/978-3-030-58583-9_8
19. He, K., Fan, H., Wu, Y., Xie, S., Girshick, R.: Momentum contrast for unsupervised visual representation learning. In: Proceedings of the IEEE/CVF Conference on Computer Vision and Pattern Recognition, pp. 9729–9738 (2020)
20. He, K., Zhang, X., Ren, S., Sun, J.: Deep residual learning for image recognition. In: Proceedings of the IEEE Conference on Computer Vision and Pattern Recognition (CVPR) (2016)
21. Hinton, G.E., Srivastava, N., Krizhevsky, A., Sutskever, I., Salakhutdinov, R.R.: Improving neural networks by preventing co-adaptation of feature detectors. arXiv preprint arXiv:1207.0580 (2012)
22. Hou, R., Chang, H., Ma, B., Shan, S., Chen, X.: Cross attention network for few-shot classification. In: Advances in Neural Information Processing Systems, vol. 32 (2019)
23. Howard, J., Ruder, S.: Universal language model fine-tuning for text classification. arXiv preprint arXiv:1801.06146 (2018)
24. Hu, S.X., Li, D., Stühmer, J., Kim, M., Hospedales, T.M.: Pushing the limits of simple pipelines for few-shot learning: external data and fine-tuning make a difference. In: Proceedings of the IEEE/CVF Conference on Computer Vision and Pattern Recognition, pp. 9068–9077 (2022)
25. Huang, Z., Wang, H., Xing, E.P., Huang, D.: Self-challenging improves cross-domain generalization. In: Vedaldi, A., Bischof, H., Brox, T., Frahm, J.-M. (eds.) ECCV 2020, Part II. LNCS, vol. 12347, pp. 124–140. Springer, Cham (2020). https://doi.org/10.1007/978-3-030-58536-5_8
26. Jing, L., Tian, Y.: Self-supervised visual feature learning with deep neural networks: a survey. IEEE Trans. Pattern Anal. Mach. Intell. **43**(11), 4037–4058 (2020)
27. Khan, S., Naseer, M., Hayat, M., Zamir, S.W., Khan, F.S., Shah, M.: Transformers in vision: a survey. ACM Comput. Surv. (CSUR) **54**(10s), 1–41 (2022)
28. Kim, J., Kim, T., Kim, S., Yoo, C.D.: Edge-labeling graph neural network for few-shot learning. In: Proceedings of the IEEE/CVF Conference on Computer Vision and Pattern Recognition, pp. 11–20 (2019)
29. Lake, B.M., Salakhutdinov, R., Tenenbaum, J.B.: Human-level concept learning through probabilistic program induction. Science **350**(6266), 1332–1338 (2015)
30. LeCun, Y., Bengio, Y., Hinton, G.: Deep learning. Nature **521**(7553), 436–444 (2015)
31. Liu, B., Zhao, Z., Li, Z., Jiang, J., Guo, Y., Ye, J.: Feature transformation ensemble model with batch spectral regularization for cross-domain few-shot classification. arXiv preprint arXiv:2005.08463 (2020)
32. Liu, L., Hamilton, W., Long, G., Jiang, J., Larochelle, H.: A universal representation transformer layer for few-shot image classification. arXiv preprint arXiv:2006.11702 (2020)
33. Liu, Q., Cao, W.: Geometric algebra graph neural network for cross-domain few-shot classification. Appl. Intell. **52**(11), 12422–12435 (2022)
34. Mazumder, P., Singh, P., Namboodiri, V.P.: Few-shot image classification with composite rotation based self-supervised auxiliary task. Neurocomputing **489**, 179–195 (2022)

35. Oh, J., Kim, S., Ho, N., Kim, J.H., Song, H., Yun, S.Y.: Refine: re-randomization before fine-tuning for cross-domain few-shot learning. In: Proceedings of the 31st ACM International Conference on Information & Knowledge Management, pp. 4359–4363 (2022)
36. Oord, A.V.D., Li, Y., Vinyals, O.: Representation learning with contrastive predictive coding. arXiv preprint arXiv:1807.03748 (2018)
37. O'Shea, K., Nash, R.: An introduction to convolutional neural networks. arXiv preprint arXiv:1511.08458 (2015)
38. Papoulis, A., Unnikrishna Pillai, S.: Probability, random variables and stochastic processes (2002)
39. Rajeswaran, A., Finn, C., Kakade, S.M., Levine, S.: Meta-learning with implicit gradients. In: Advances in Neural Information Processing Systems, vol. 32 (2019)
40. Ren, J., Li, M., Liu, Z., Zhang, Q.: Disentanglement, visualization and analysis of complex features in DNNs (2020)
41. Sa, L., Yu, C., Ma, X., Zhao, X., Xie, T.: Attentive fine-grained recognition for cross-domain few-shot classification. Neural Comput. Appl. **34**(6), 4733–4746 (2022)
42. Sagi, O., Rokach, L.: Ensemble learning: a survey. Wiley Interdiscipl. Rev. Data Min. Knowl. Discov. **8**(4), e1249 (2018)
43. Satorras, V.G., Estrach, J.B.: Few-shot learning with graph neural networks. In: International Conference on Learning Representations (2018)
44. Scarselli, F., Gori, M., Tsoi, A.C., Hagenbuchner, M., Monfardini, G.: The graph neural network model. IEEE Trans. Neural Networks **20**(1), 61–80 (2008)
45. Shorten, C., Khoshgoftaar, T.M.: A survey on image data augmentation for deep learning. J. Big Data **6**(1), 1–48 (2019)
46. Sun, J., Cao, X., Liang, H., Huang, W., Chen, Z., Li, Z.: New interpretations of normalization methods in deep learning. In: Proceedings of the AAAI Conference on Artificial Intelligence, vol. 34, pp. 5875–5882 (2020)
47. Torralba, A., Efros, A.A.: Unbiased look at dataset bias. In: CVPR 2011, pp. 1521–1528. IEEE (2011)
48. Triantafillou, E., et al.: Meta-dataset: a dataset of datasets for learning to learn from few examples. arXiv preprint arXiv:1903.03096 (2019)
49. Tseng, H.Y., Lee, H.Y., Huang, J.B., Yang, M.H.: Cross-domain few-shot classification via learned feature-wise transformation. arXiv preprint arXiv:2001.08735 (2020)
50. Tu, P.C., Pao, H.K.: A dropout style model augmentation for cross domain few-shot learning. In: 2021 IEEE International Conference on Big Data (Big Data), pp. 1138–1147. IEEE (2021)
51. Veilleux, O., Boudiaf, M., Piantanida, P., Ben Ayed, I.: Realistic evaluation of transductive few-shot learning. Adv. Neural. Inf. Process. Syst. **34**, 9290–9302 (2021)
52. Vinyals, O., Blundell, C., Lillicrap, T., Wierstra, D., et al.: Matching networks for one shot learning. In: Advances in Neural Information Processing Systems, vol. 29 (2016)
53. Wah, C., Branson, S., Welinder, P., Perona, P., Belongie, S.: The caltech-ucsd birds-200-2011 dataset (2011)
54. Wang, F., et al.: Residual attention network for image classification. In: Proceedings of the IEEE Conference on Computer Vision and Pattern Recognition, pp. 3156–3164 (2017)

55. Wang, H., et al.: Experiments in cross-domain few-shot learning for image classification. In: ECMLPKDD Workshop on Meta-Knowledge Transfer, pp. 81–83. PMLR (2022)
56. Wang, M., Deng, W.: Deep visual domain adaptation: a survey. Neurocomputing **312**, 135–153 (2018)
57. Wang, Y., Yao, Q., Kwok, J.T., Ni, L.M.: Generalizing from a few examples: a survey on few-shot learning. ACM Comput. Surv. (CSUR) **53**(3), 1–34 (2020)
58. Xu, Y., Wang, L., Wang, Y., Qin, C., Zhang, Y., Fu, Y.: Memrein: rein the domain shift for cross-domain few-shot learning (2021)
59. Yang, L., Li, L., Zhang, Z., Zhou, X., Zhou, E., Liu, Y.: DPGN: distribution propagation graph network for few-shot learning. In: Proceedings of the IEEE/CVF Conference on Computer Vision and Pattern Recognition, pp. 13390–13399 (2020)
60. Yazdanpanah, M., Moradi, P.: Visual domain bridge: a source-free domain adaptation for cross-domain few-shot learning. In: Proceedings of the IEEE/CVF Conference on Computer Vision and Pattern Recognition, pp. 2868–2877 (2022)
61. Yazdanpanah, M., Rahman, A.A., Desrosiers, C., Havaei, M., Belilovsky, E., Kahou, S.E.: Shift and scale is detrimental to few-shot transfer. In: NeurIPS 2021 Workshop on Distribution Shifts: Connecting Methods and Applications (2021)
62. Ye, H.J., Hu, H., Zhan, D.C., Sha, F.: Few-shot learning via embedding adaptation with set-to-set functions. In: Proceedings of the IEEE/CVF Conference on Computer Vision and Pattern Recognition, pp. 8808–8817 (2020)
63. Zhang, H., Cisse, M., Dauphin, Y.N., Lopez-Paz, D.: mixup: beyond empirical risk minimization. arXiv preprint arXiv:1710.09412 (2017)
64. Zhang, Q., Jiang, Y., Wen, Z.: TACDFSL: task adaptive cross domain few-shot learning. Symmetry **14**(6), 1097 (2022)
65. Zhang, T., Wu, F., Katiyar, A., Weinberger, K.Q., Artzi, Y.: Revisiting few-sample BERT fine-tuning. arXiv preprint arXiv:2006.05987 (2020)
66. Zhang, Y., Zheng, Y., Xu, X., Wang, J.: How well do self-supervised methods perform in cross-domain few-shot learning? arXiv preprint arXiv:2202.09014 (2022)
67. Zhang, Y., Li, W., Zhang, M., Tao, R.: Dual graph cross-domain few-shot learning for hyperspectral image classification. In: ICASSP 2022–2022 IEEE International Conference on Acoustics, Speech and Signal Processing (ICASSP), pp. 3573–3577. IEEE (2022)
68. Zhang, Y., Li, W., Zhang, M., Wang, S., Tao, R., Du, Q.: Graph information aggregation cross-domain few-shot learning for hyperspectral image classification. IEEE Trans. Neural Networks Learn. Syst. (2022)
69. Zhao, Y., Gao, X., Shumailov, I., Fusi, N., Mullins, R.: Rapid model architecture adaption for meta-learning. Adv. Neural. Inf. Process. Syst. **35**, 18721–18732 (2022)
70. Zhou, K., Liu, Z., Qiao, Y., Xiang, T., Loy, C.C.: Domain generalization: a survey. IEEE Trans. Pattern Anal. Mach. Intell. **45**, 4396–4415 (2022)

Exploring the Predictive Power of Correlation and Mutual Information in Attention Temporal Graph Convolutional Network for COVID-19 Forecasting

Subas Rana(✉), Nasid Habib Barna, and John A. Miller

University of Georgia, Athens, GA 30602, USA
{subas.rana,nasidhabib.barna,jamill}@uga.edu

Abstract. Accurately forecasting the COVID-19 spread across all states is crucial for implementing effective measures to control its transmission and minimize its impact. Since the virus' spread in one state can significantly affect other states over time through connections between them, a graph structure with temporal data is needed to capture the interdependence of COVID-19 spread among the states in the United States. In forecasting tasks that involve complex spatial and temporal dependencies, it is crucial to ensure that the model captures these dependencies accurately. In this study, we implemented an Attention Temporal Graph Convolutional Network based model for COVID-19 mortality long-term prediction which can effectively capture these dependencies. This model incorporates attention that enables us to weigh the significance of different time points and focus on the most informative data, including both adjacent and distant time points that capture the temporal dynamics accurately. For capturing spatial dependencies, we assessed the impact of using Pearson's correlation and Mutual Information to establish connections between highly dependent states. Our experiments showed that our model, particularly when utilizing mutual information, outperformed the existing baselines and the models that only consider neighboring states resulting in lower sMAPE and MAE values. This emphasizes the importance of selecting the appropriate technique for accurate COVID-19 forecasting in each state. Furthermore, our model achieved the second-highest performance among the forecasting models submitted to the Centers for Disease Control and Prevention.

Keywords: COVID-19 forecasting · Attention Temporal Graph Convolutional Network · PyTorch Geometric Temporal · Pearson's Correlation · Mutual Information

Supported by University of Georgia.

S. Zhang et al. (Eds.): BigData 2023, LNCS 14203, pp. 18–33, 2023.
https://doi.org/10.1007/978-3-031-44725-9_2

1 Introduction

The first case of COVID-19 in the United States (US) was reported in the state of Washington on January 20, 2020. Since then, the virus has spread rapidly across all 50 states, resulting in over 104 million confirmed cases and 1 million deaths as of May 4, 2023 [1]. The availability of daily and weekly COVID-19 data from various countries and states has provided opportunities to develop improved time-series models. Several statistical and machine learning (ML) models [2–4] have been used for COVID-19 forecasting. These methods either use the epidemic data of a single region to predict the future trend of the epidemic situation or establish a generalized model to predict the trend of all regions. Extensive research [5,6] has confirmed the highly contagious nature of the virus, and these studies have identified factors that contribute to its rapid spread across regions. Consequently, the spread of COVID-19 within a particular state can exert a substantial influence on neighboring states over time. This influence arises from various interconnected factors such as travel, trade, and social connections. To accurately forecast the spread of the virus in a particular state, it is important to consider the graph structure to capture these interconnections between states.

Recent studies [7–9] have developed spatiotemporal Graph Neural Networks (GNNs) that operate on graphs, representing regions as nodes and capturing spatial and temporal dependencies between them. In forecasting tasks that involve complex spatial and temporal dependencies, it is crucial to ensure that the model captures both types of relationships accurately. These GNNs combine Graph Convolutional Networks (GCNs) to capture spatial dependencies and Recurrent Neural Networks (RNNs) or their variants to model temporal dependencies. RNNs process sequential data over time, and their hidden states carry the latest information from the past. However, this sequential processing can restrict the model's access to global information present throughout the input sequence. To address this issue, Attention mechanisms offer a solution by allowing the model to focus on relevant parts of the sequence and assign different weights, providing a means to learn and leverage global correlations. We utilize an attention-based model from PyTorch Geometric Temporal library (PyGT) [10] to capture global information for COVID-19 weekly mortality prediction.

Another crucial aspect of GNN-based models is to accurately capture strong connections between different regions. Many existing GNN-based models [9,11,12] for epidemic prediction rely on data such as mobility or social connections to establish connections between regions and capture spatial dependencies. However, acquiring and utilizing such data can pose challenges related to data availability, privacy, and accuracy. In contrast, we utilize both correlation and mutual information (MI) to capture both linear and nonlinear dependencies between the state's features. For instance, our approach shows a high correlation between Ohio and Illinois, which suggests there is a strong linear relationship between the deaths or confirmed cases in these two states. Specifically, it suggests that as deaths or confirmed cases increase in one state, they tend to increase in the other state as well. This relationship can be seen in Fig. 1.

The contributions of our proposed method are as follows:

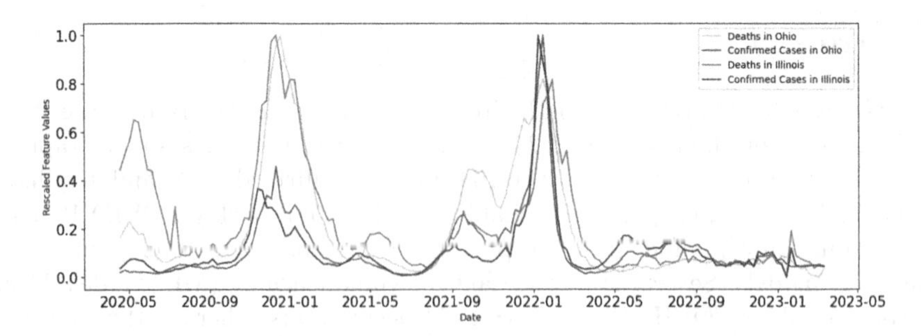

Fig. 1. Rescaled Deaths and Confirmed Cases in Ohio and Illinois indicating a strong linear relationship

1. We use the Attention Temporal Graph Convolutional Network (A3T-GCN) model from PyGT on the state graph for COVID-19 forecasting. The complete model allowed us to capture both local and global information about all the regions in the model, allowing us to capture temporal dependencies effectively.
2. We use correlation and MI to capture both linear and nonlinear dependencies and establish connections between highly dependent states in a graph, allowing us to effectively capture spatial dependencies. We use these graphs in the A3T-GCN model and compare these techniques with A3T-GCN based on adjacent states (states that share a border). We also compare our best models with the baselines.
3. We compared and achieved the second-highest performance among the forecasting models submitted to the Centers for Disease Control and Prevention (CDC).

2 Related Work

Several studies have applied statistical, ML, and deep learning methods to predict COVID-19 forecasting based on various clinical, laboratory, and epidemiological data. Infectious disease prediction is often modeled as a time-series prediction problem in many studies, so many time-series methods have been studied [3]. Chimmula et al. [2] analyzed the important factors and applied a Long Short Term Memory (LSTM) model to forecast the patterns and probable end date of the COVID-19 pandemic in Canada and also around the world. Cramer et al. [13] conducted a study that evaluates the effectiveness of both individual and ensemble probabilistic forecasts for predicting COVID-19 mortality in the US. The study focuses on evaluating the accuracy and reliability of these forecasts to provide insights into the effectiveness of different forecasting methods. Their work has been used by CDC for COVID-19 cases and death forecasts.

Kapoor et al. [7] presented a forecasting method for COVID-19 cases using a spatiotemporal GNN that aims to capture the intricate dynamics involved in disease modeling by incorporating mobility data. In their proposed model, nodes in the graph represent county-level human mobility, spatial edges depict inter-regional interactions, and temporal edges account for the evolution of node features over time. Panagopoulos et al. [9] introduced MPNN-TL, a GNN for COVID-19 dynamics across four European countries. By integrating mobility data and reported cases, the model aims to comprehend complex transmission patterns. It recognizes the vital role of mobility patterns in the disease's spread, emphasizing the significance of a graph-based representation for studying transmission dynamics and its impacts. Fritz et al. [14] proposed a fusion approach that combines GNN with epidemiological models to forecast the infection rate of COVID-19. The study utilized data from Facebook, including mobility and association information, as well as structural and spatial details of cities and districts in Germany. Cao et al. [15] developed StemGNN, a model for multivariate time series that captures inter and intra-temporal correlations using the Graph Fourier Transform (GFT) and Discrete Fourier Transform (DFT). With a graph structure representing different countries, the study evaluated the model's performance in forecasting confirmed cases across multiple horizons.

3 Methodology

In this work, we utilized the A3T-GCN model from PyGT on the state graph to predict COVID-19 outcomes. This model allowed us to effectively consider both local and global information from all regions, capturing temporal dependencies. To capture both linear and nonlinear dependencies and establish connections between highly dependent states, we incorporated correlation and MI. By integrating these graphs into the A3T-GCN model, we compared its performance with the version that only considered adjacent states. Our evaluation showed a significant improvement when using different association techniques beyond adjacency. In this section, we will discuss the PyGT library, the model architecture with a customized dataset, and the association techniques used for the model.

3.1 PyGT

PyGT is an extension library for PyTorch Geometric. It is specifically designed for handling spatiotemporal data, such as dynamic graphs where edges and nodes change over time. PyGT includes various tools for creating, manipulating, and visualizing temporal graphs, as well as implementing GNN architectures for spatiotemporal data. This library provides several data iterators. We use **StaticGraphTemporalSignal** which is used when the underlying graph is fixed, but the features on each node or edge change over time. This data iterator provides an efficient way to iterate over temporal snapshots of a graph in batches. The library also comes with a train-test splitter that creates temporal splits of the data using a fixed split ratio and some benchmark datasets. In addition to that, it includes several types of existing Neural network models that operate on graphs. We use the model A3T-GCN which is based on the paper "A3T-GCN: Attention Temporal Graph Convolutional Network for Traffic Forecasting" [16].

3.2 A3T-GCN

We implement an A3T-GCN (the second version of A3T-GCN) based model which is capable of handling batches. Our objective with this model is to use past values of COVID-19 deaths and confirmed cases to predict the weekly death counts for each state. This model, shown in Fig. 2, is a combination of GCN and Gated Recurrent Unit (GRU) and features an attention mechanism.

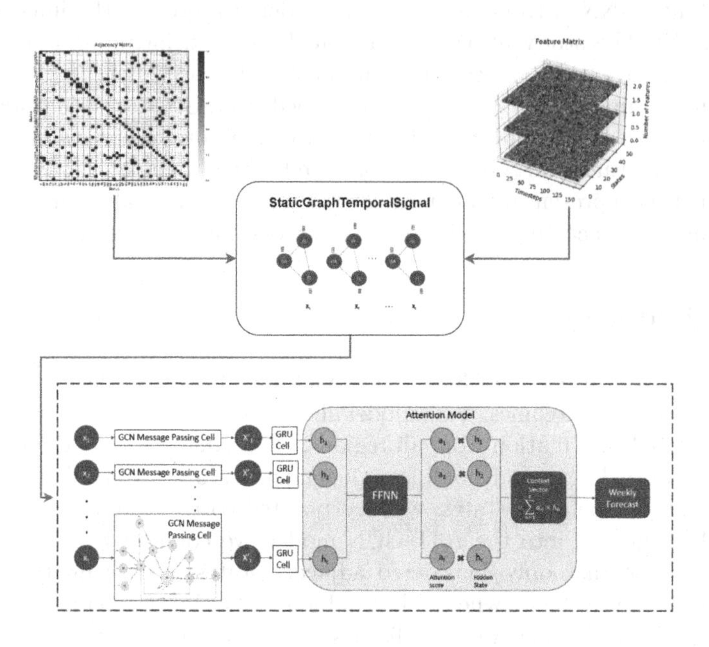

Fig. 2. Customized PyGT dataset based A3T-GCN architecture

The library requires a customized PyTorch Geometric-based dataset that is tailored to the problem of predicting COVID-19 deaths for the states. To achieve this, we define an adjacency matrix (based on adjacent states as an example in this figure) and a feature matrix with information on timesteps, number of states, and features. These matrices are passed to the model expanded from the A3T-GCN [16] paper as shown in Fig. 2.

1. **Adjacency Matrix:** A graph $G = (V, E)$ defines the graph structure of the US states where each node $v \in V$ represents a state in the US and each edge $e \in E$ represents the edge between two states. This whole information is defined by an adjacency matrix $A \in R^{N \times N}$ with N vertices (states in our case) and all the rows i and columns j are indexed by the states. The entry in i and j of the matrix, denoted by $A[i, j]$, represents the weight or degree of interdependence between state i and state j.
   We use correlation and MI described in the next section to construct the matrices based on the degree of dependency between pairs of states. The

procedure involves calculating the correlation and MI scores between the variables. We consider all pairs of states with a score greater than a certain threshold to be highly dependent and connect them in the adjacency matrix. Conversely, we set the weight of the edge to zero for all pairs of states with the scores below the threshold, indicating that they are not strongly interconnected.

2. **Feature Matrix:** COVID-19 deaths and confirmed cases are shown as the features of the nodes represented by a feature matrix $X \in R^{N \times F}$, where N is the number of states and F is the number of features. These features change over time, X_t is the features matrix at time t. We passed the historical values of all the features $X_{t-n}, \cdot, X_{t-1}, X_t$ along with the adjacency matrix A through the **StaticGraphTemporalSignal** iterator which returns PyTorch Geometric Data object for a single time period i.e., a week, which represents a snapshot of the graph for that time period. These temporal snapshots represent different feature values for the same underlying graph structure and are passed as historical inputs to the GCN layers of the A3T-GCN model, which perform computations on the graph structure and associated features to produce output representations of the hidden state.

GCN. In a GCN, each node in the graph is associated with a feature vector, and the goal is to learn a new set of feature vectors that capture the relationships in the graph through a message-passing process. Specifically, at each layer of the GCN, the feature vector of each node is updated by taking a weighted sum of the feature vectors of its neighbors, where the weights are learned by the network. This weighted sum is then combined with the feature vector of the node itself to produce a new feature vector. This enables the model to capture spatial dependencies. A two-layer GCN model [17] can be defined in (1) below:

$$f(X, A) = \sigma(\overline{A}\ Sigmoid\ (\overline{A}\ X\ W_0)\ W_1) \tag{1}$$

$$\overline{A} = D'^{-.5} A' D'^{-.5} \tag{2}$$

$$A' = A + I_N \tag{3}$$

Here X is a feature matrix and A is an adjacency matrix defined above, $\overline{A}$ defined in (2) is a preprocessing step, where A' is an adjacency matrix with self connectivity defined in (3), I_N is an identity matrix, D' is a degree matrix. $W_0 \in R^{K \times H}$ is a weight matrix from the input layer to the hidden unit layer where K is the length of time and H is hidden unit numbers, and $W_1 \in R^{H \times T}$ is a weight matrix from the hidden layer to the output layer, where T is the length of forecasted points. $f(X, A) \in R^{N \times T}$ is the output of the GCN model of the length T. The updated vectors are then passed to GRU.

GRU. GRUs have two gates, namely an update gate and a reset gate. The update gate determines how much of the previous hidden state should be retained for the current time step, and the reset gate controls how much of the new input should be incorporated into the new hidden state.

In our case, the gated mechanism of GRU allows them to capture the mortality information at the current moment while retaining the variation trends of historical COVID information. As a result, this model can effectively capture the dynamic temporal variation features of COVID data.

Attention. The Attention model, which is a modified version of the Encoder-Decoder model, was initially developed for use in neural machine translation tasks [18]. In this particular study, a soft Attention model was employed to determine the importance of COVID information at each moment. This information was then used to calculate a context vector that captured the overall trends in the COVID mortality state, which could be utilized for predicting future mortality conditions.

In order to calculate the weight of each hidden state, a scoring function is designed, followed by an attention function that computes the context vector to capture the global COVID data variation information.

$$e_i = W_2(W_1 H + d_1) + d_2 \tag{4}$$

$$a_i = \frac{exp(e_i)}{\sum_{k=1}^{n} exp(e_k)} \tag{5}$$

$$C_t = \sum_{i=1}^{n} a_i \times h_i \tag{6}$$

Here H is the set of hidden states $\{h_1, h_2, \ldots, h_n\}$, W_1 and d_1 are the weight and deviation of the first layer and W_2 and d_2 are the weight and deviation of the second layer. The context vector $C_t \in R^{N \times T}$ captures global COVID data information and lastly, a linear layer is used to produce the final output results. In this study, a Feed Forward Neural Network (FFNN) was used instead of a scoring function to determine the weight of each hidden state resulting from the GRU.

3.3 Correlation

Correlation is used to measure the linear relationships between the variables. However, it does not measure the strength of a nonlinear relationship between variables. Figure 3a shows the adjacency matrix based on correlation. To connect the states/nodes, we only selected those with high correlation coefficients.

3.4 MI

MI [19] measures the mutual dependence between two random variables. It measures the amount of information that one random variable provides about the other random variable. It can measure both linear and nonlinear relationships between two random variables. MI is derived from the definition of entropy and is defined as:

$$I(X;Y) = \sum_{x \in X} \sum_{y \in Y} P(x,y) \log \frac{P(x,y)}{P(x)P(y)} \tag{7}$$

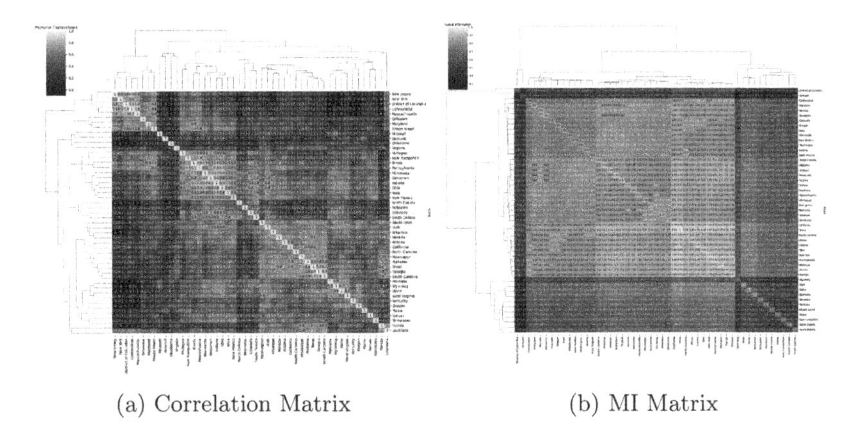

(a) Correlation Matrix (b) MI Matrix

Fig. 3. Association matrices

Let X and Y be two random variables over the space $X \times Y$. $I(X;Y)$ is the MI calculated between the two variables, $P(X,Y)$ is their joint distribution and $P(X)$ and $P(Y)$ is the marginal distribution of X and Y, respectively. The MI score is greater than or equal to zero. If it is zero, it means that the two variables are independent, i.e., knowing the value of one variable does not provide any information about the other variable. If the MI score is greater than zero, the two variables are dependent, and the larger the MI score, the stronger the dependence.

Figure 3b shows the MI-based matrix. To create edges between the states, we only selected those with high MI scores. Additionally, we rescaled the MI values by dividing each value by its maximum.

For both techniques, we experimented with different thresholds and selected the ones that yielded the best results. We used 0.6 for correlation and 0.85 for MI which preserved the strong connections between states while discarding weaker ones that fell below the threshold. For creating a weighted graph, we assigned the scores as edge weights for connected edges and assigned 0 for disconnected ones. Figure 4 shows how the graph changes based on the association techniques.

4 Experiments and Results

4.1 Data and Code

The weekly dataset used for this study is available on GitHub https://github.com/scalation/data/blob/master/COVID-State/2023-05-02-17-53-19-State_Weekly.csv. It contains 19 columns and 8819 rows corresponding to weekly reporting for each region in the US from April 18th, 2020, to March 11th, 2023. We only consider the data for states and drop other regions. For features, we use *Confirmed* and *Deaths* because other features have missing values.

The data was extracted from the COVID-19 Dataset by COVID-19 Data Repository by the Center for Systems Science and Engineering (CSSE) at

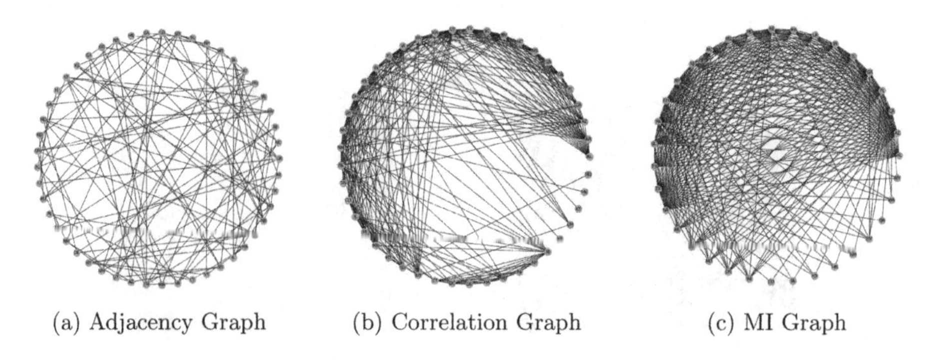

(a) Adjacency Graph (b) Correlation Graph (c) MI Graph

Fig. 4. Connection between states with different association techniques

Johns Hopkins University [20] (JHU) https://github.com/CSSEGISandData/COVID-19. Daily data were cumulative and produced some negative values when converted to non-cumulative, for instance, Maryland's dataset showed 1140 deaths on 04/30/2020 and 1080 on 05/01/2020 which resulted in -60 deaths on May 1st, 2020. To resolve this issue, we opted to aggregate the data on a weekly basis instead. Once converted, we removed the initial missing values and divided the dataset into an 80/20 split using the PyGT Temporal Signal Train-Test Split for training and testing, which returns train and test data iterators. We employed the sliding window approach for model training, where each window consisted of 114 weeks in the training set and forecasted 4 weeks in the testing set. We then shifted the window by 1 week and repeated the process until the end of the dataset. Moreover, for CDC comparison, the data was extracted from https://covid.cdc.gov/COVID-DATA-TRACKER/?submenu-select=national-lab#datatracker-home. Our code is available on: https://www.github.com/Subasranaa/COVID-19-A3T-GCN2.

4.2 Implementation Details

We implemented our model in Python using PyTorch [21] and PyGT. The hyperparameters shown in Table 1 are fine-tuned using grid search. The activation function is ReLU, 100 epochs, and a 1e-4 learning rate using the ADAM optimizer. We used the mean squared error (MSE) as the loss function. The data were standardized (subtracting the mean and dividing by the standard deviation) and transformed back to the original data scale for evaluation. The model was evaluated on the symmetric mean absolute percentage error (sMAPE) and the mean absolute error (MAE). sMAPE measures the absolute difference between the actual and forecast values normalized by absolute values of both [22]. It is a bounded metric where 0% indicates a perfect model with no errors, while 200% indicates an erroneous model of opposite actual and forecast values [23]. The sMAPE is calculated using (8)

$$\text{sMAPE} = \frac{200}{n} \sum_{t=1}^{n} \frac{|y_t - \hat{y}_t|}{|y_t| + |\hat{y}_t|} \tag{8}$$

where y_t is the true value, and $\hat{y}_t$ is the forecast value. MAE measures the mean absolute error between the actual and forecast values.

Statistical and ML Baselines. To compare our A3T-GCN model, we employed several baselines. Auto-Regressive(1) (AR) model is sufficient when the future is mainly dependent only on the most recent value. SARIMA is a time series forecasting model that combines AR, integrated (I), and moving average (MA) components to capture the patterns and seasonality in data. LSTM is designed to process and retain information over long sequences, enabling it to capture and learn from long-term dependencies in data. It uses a gating mechanism to selectively remember or forget information based on relevance. GRU is similar to LSTM but has fewer gates, making it faster to compute and easier to train. FFNN is where information flows in one direction, from the input layer through hidden layer(s) to the output layer, without any loops. We experimented with these models using various hyperparameters as shown in Table 1.

Table 1. Optimal hyperparameters for A3T-GCN and all the baselines. A3T-GCN, LSTM, GRU, and FFNN hyperparameters are listed in the sequence of [past values, batch size, hidden dimensions, learning rate]. FFNN utilized ELU and Tanh activation functions and four linear layers.

	Virginia	Georgia	Illinois	Pennsylvania	Kentucky
A3T-GCN	[6,16,4,0.0001]	[6,16,2,0.0001]	[6,16,32,0.0001]	[6,16,2,0.0001]	[6,16,16,0.0001]
SARIMA	$(6,0,1) \times (0,1,1)_{10}$	$(3,1,2) \times (1,0,1)_{10}$	$(0,1,0) \times (1,1,1)_{10}$	$(4,0,0) \times (5,1,1)_{10}$	$(3,1,1) \times (1,1,1)_{10}$
LSTM	[6,16,256,0.0006]	[6,16,32,0.0006]	[6,16,128,0.0006]	[6,16,256,0.0006]	[6,16,512,0.0006]
GRU	[6,16,128,0.0007]	[6,16,512,0.0007]	[6,16,64,0.0007]	[6,16,256,0.0007]	[6,16,128,0.0007]
FFNN	[4,16,256,0.0004]	[4,16,32,0.0004]	[4,16,32,0.0004]	[4,16,16,0.0004]	[4,16,512,0.0004]
	Maryland	Massachusetts	Minnesota	Ohio	Washington
A3T-GCN	[6,16,8,0.0001]	[6,16,32,0.0001]	[6,4,32,0.0001]	[6,16,16,0.0001]	[6,16,32,0.0001]
SARIMA	$(0,1,2) \times (1,0,1)_{10}$	$(1,1,0) \times (0,0,2)_{10}$	$(1,1,1) \times (0,0,1)_{10}$	$(2,0,0) \times (0,1,2)_{10}$	$(4,0,1) \times (0,0,1)_{10}$
LSTM	[6,16,8,0.0006]	[6,16,128,0.0006]	[6,16,64,0.0006]	[6,16,256,0.0006]	[6,16,512,0.0006]
GRU	[6,16,512,0.0007]	[6,16,32,0.0007]	[6,16,4,0.0007]	[6,16,512,0.0007]	[6,16,256,0.0007]
FFNN	[4,16,16,0.0004]	[4,16,32,0.0004]	[4,16,32,0.0004]	[4,16,16,0.0004]	[4,16,16,0.0004]

4.3 Results Analysis

Comparison of A3T-GCN Models Using Adjacent, Correlation, and MI Matrices. We implemented five A3T-GCN models using Adjacent, Correlation, MI, Correlation with adjacent (Corr_Adj), and MI with adjacent (MI_Adj) matrices. The reason behind this is each state was showing different trends when they were connected to different states based on their relation. We are using the adjacent-based A3T-GCN model as the baseline here to compare it with other association technique-based models. Tables 2 and 3 show sMAPE and MAE results respectively, of 1 week, 2 weeks, 3 weeks, and 4 weeks ahead forecast for our best models and the baseline in this case.

Virginia, Ohio, Pennsylvania, and Washington demonstrate improved outcomes when using the MI based A3T-GCN model, suggesting that quantifying the interdependence among these states through MI yields favorable results. The MI_Adj-based model gives us the best results for Georgia, Kentucky, Massachusetts, and Minnesota indicating that the involvement of neighboring states is significant for these particular states. For instance, in the case of Georgia, 21 more states got added along with adjacent states using MI which improved the predictions, showing a strong impact of linear and nonlinear dependencies with other states.

In the case of Maryland and Illinois, the Corr_Adj-based model shows superior performance than all the other models indicating that correlation plays an important role in them. Here adding correlated states led to superior results instead of just using adjacent states, our models give a slightly higher MAE score for Maryland even though it has a lower sMAPE score when we compare it with the adjacent-based model. Moreover, we also get competitive scores for Arizona, Iowa, Michigan, New York, Texas, and Indiana with sMAPEs as 25.52, 28.74, 26.87, 26.90, 25.25, and 27.53, respectively by using MI. Overall, using these association techniques over using only adjacent states shows significantly better performance in terms of forecasting.

Table 2. sMAPE results for 1-week, 2-weeks, 3-weeks, and 4-weeks ahead forecast for the states using Adjacent states-based A3T-GCN (Adjacent) and Correlation/MI-based A3T-GCN models (A3T-GCN). The lower the sMAPEs, the better the forecasting performance.

	Virginia		Georgia		Illinois		Pennsylvania		Kentucky	
Weeks	Adjacent	A3T-GCN	Adjacent	A3T-GCN	Adjacent	A3T-GCN	Adjacent	A3T-GCN	Adjacent	A3T-GCN
1	39.96	13.02	22.82	15.07	23.12	11.38	88.93	17.14	20.74	14.28
2	45.36	12.58	27.50	13.77	21.57	9.37	90.10	28.46	19.14	12.37
3	48.41	11.14	33.24	17.77	16.98	10.81	91.32	15.89	16.53	11.23
4	53.60	13.55	22.27	26.38	19.93	14.66	95.40	23.07	14.75	11.83
Average	18.36	**12.57**	26.46	**18.25**	20.40	**11.56**	47.24	**21.14**	17.79	**12.43**
	Maryland		Massachusetts		Minnesota		Ohio		Washington	
Weeks	Adjacent	A3T-GCN	Adjacent	A3T-GCN	Adjacent	A3T-GCN	Adjacent	A3T-GCN	Adjacent	A3T-GCN
1	23.47	14.11	15.44	13.65	18.38	12.31	16.16	14.81	28.27	12.90
2	18.87	14.34	17.16	13.91	14.10	12.04	18.16	6.52	27.31	13.95
3	14.45	18.04	18.09	13.33	21.43	14.35	16.26	7.28	30.05	13.90
4	15.83	23.34	20.84	13.45	19.45	13.94	16.59	14.63	31.79	12.77
Average	18.16	**17.46**	17.88	**13.58**	18.34	**13.16**	16.79	**10.81**	29.35	**13.38**

Table 3. MAE results for 1-week, 2-weeks, 3-weeks, and 4-weeks ahead forecast for the states using Adjacent states-based A3T-GCN (Adjacent) and Correlation/MI-based A3T-GCN models (A3T-GCN). The lower the MAEs, the better the forecasting performance.

	Virginia		Georgia		Illinois		Pennsylvania		Kentucky	
Weeks	Adjacent	A3T-GCN	Adjacent	A3T-GCN	Adjacent	A3T-GCN	Adjacent	A3T-GCN	Adjacent	A3T-GCN
1	32.76	12.36	27.10	17.38	18.55	9.11	46.32	19.92	12.94	10.35
2	36.37	11.84	34.41	15.76	16.05	7.27	47.59	30.70	11.78	8.65
3	38.81	10.61	43.93	20.69	12.78	8.52	48.63	18.90	10.91	7.86
4	42.57	12.81	26.31	32.69	15.49	11.64	52.66	29.60	10.68	8.27
Average	37.63	**11.90**	32.94	**21.63**	15.72	**9.13**	48.80	**24.78**	11.58	**8.78**
	Maryland		Massachusetts		Minnesota		Ohio		Washington	
Weeks	Adjacent	A3T-GCN	Adjacent	A3T-GCN	Adjacent	A3T-GCN	Adjacent	A3T-GCN	Adjacent	A3T-GCN
1	9.53	5.75	14.46	8.70	7.70	5.99	13.99	11.34	19.11	8.62
2	7.75	5.90	15.65	8.87	6.14	5.83	15.63	5.14	18.33	9.32
3	5.81	7.90	15.90	8.47	9.09	7.00	14.20	5.85	20.69	9.22
4	6.62	10.71	19.25	8.56	9.20	6.83	14.48	12.14	22.20	8.52
Average	**7.43**	7.57	16.32	**8.65**	8.03	**6.41**	14.57	**8.62**	20.08	**8.92**

Baselines Comparison. In this work, we compare our best-performing A3T-GCN models against the baselines. Figure 5 displays the average results of sMAPE and MAE for our best A3T-GCN models and the baselines across the states. A3T-GCN achieves a 52.96% reduction in sMAPE and a 46.84% reduction in MAE compared to AR and a 37.99% reduction in sMAPE and a 36.32% reduction in MAE compared to SARIMA. Although LSTM performs slightly better than A3T-GCN for Maryland, when considering the average results across all states, A3T-GCN demonstrates a 15.86% lower sMAPE and a 16.26% lower MAE. Compared to GRU, A3T-GCN achieves a 22.55% lower sMAPE and a 26.72% lower MAE, despite GRU outperforming in the case of Massachusetts and Minnesota (for MAE only). Finally, when compared to FFNN, A3T-GCN shows a 35.03% lower sMAPE and a 38.51% lower MAE.

This is likely due to the model's use of Attention, GCN, and GRU, which can capture non-linear dependencies and temporal relationships between distant data points. In contrast, AR and SARIMA may not be able to capture these non-linear dependencies and relationships, leading to lower performance. The findings also highlight that relying solely on historical data from a single state may not be sufficient for accurately predicting deaths, as the interactions and relationships between different states can impact the results. Therefore, it is crucial to consider these relationships in the model, which A3T-GCN can capture. This becomes particularly significant when dealing with time series data. While LSTM, FFNN, and GRU performed well, their limitations in capturing these relationships might have hindered their performance for some states.

CDC Analysis. We compared our best A3T-GCN models with the models submitted to CDC for the states. The evaluation of these models was based on the sMAPE definition, and Table 4 shows the average sMAPE score for all states, excluding Virginia and Pennsylvania because the data for those two states

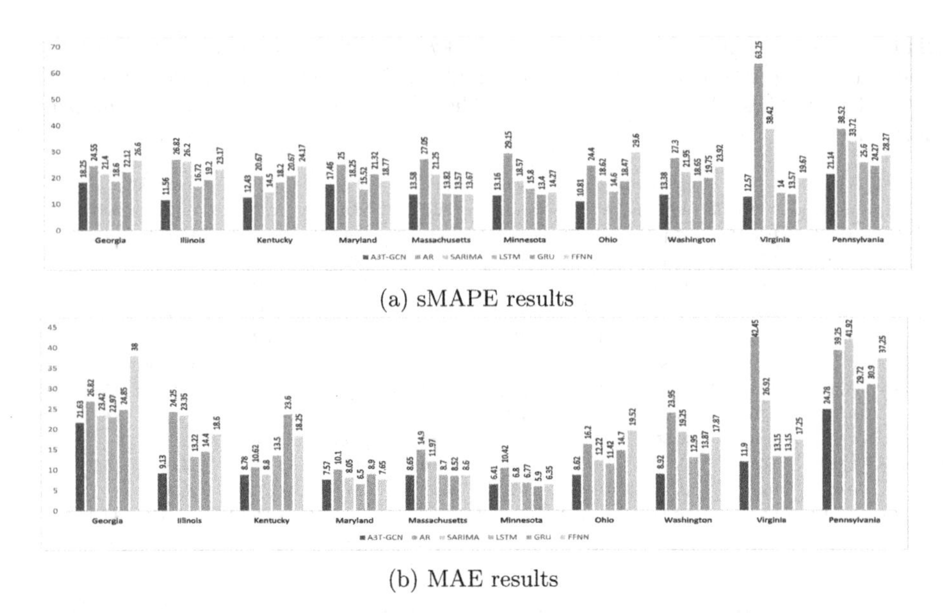

(a) sMAPE results

(b) MAE results

Fig. 5. Average of prediction results of the best A3T-GCN and the baselines

was unavailable. We limited our selection to only those models with matching forecasting dates. A3T-GCN ranked second overall, outperforming most other models. In addition, our model demonstrated superior performance for Ohio, as depicted in Fig. 6.

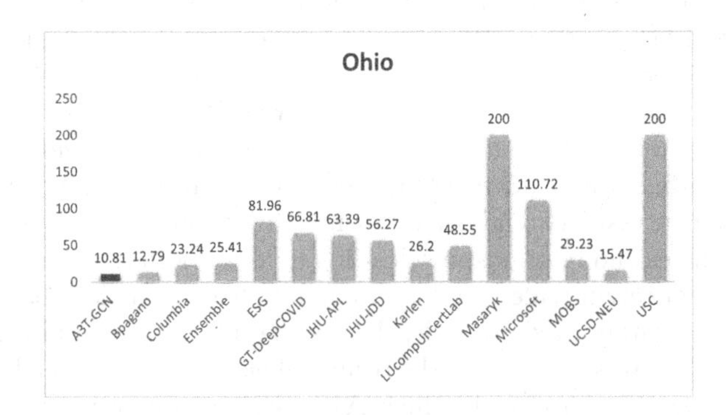

Fig. 6. A3T-GCN comparison with CDC models for Ohio using sMAPE

5 Conclusion and Future Work

Our work highlights the importance of accurately capturing both spatial and temporal dependencies in COVID-19 forecasting. The MI and correlation-based

Table 4. Comparison of A3T-GCN with the models forecast submitted to CDC using sMAPE

Model	sMAPE	Model Method
JHU-IDD [24]	82.67	Metapopulation Susceptible-Exposed-Infected-Recovered (SEIR) model
LUcompUncertLab [25]	66.33	A Bayesian Vector Auto Regression model
Microsoft [26]	53.1	SEIR model on a spatiotemporal network
USC [27]	48.82	Discrete heterogeneous rate model
ESG [28]	36.04	Fitting reported data to multiple skewed gaussian distributions
Masaryk	35.59	ARIMA models with outlier detection applied to transformed series
UCSD-NEU [29]	26.54	Age-structured metapopulation model with deep learning
JHU-APL [30]	25.73	Metapopulation SEIR model
Karlen [31]	23.33	Discrete time difference equations
MOBS [32]	21.3	Metapopulation, age-structured Susceptible-Latent-Infected-Removed (SLIR) model
GT-DeepCOVID [33]	20.22	Deep Learning
BPagano [34]	15.97	Susceptible-Infected-Recovered (SIR) model
Ensemble [35]	14.28	Combination of several forecasts
A3T-GCN	**13.82**	Our Model
Columbia [36]	10.51	Metapopulation SEIR model

A3T-GCN model proposed in this work addresses this challenge by leveraging the attention mechanism to focus on the most informative data that includes both recent and distant data points to capture temporal dependencies and correlation and MI to capture the spatial dependencies effectively. We found utilizing MI to be the most effective technique to create the graph that the model uses for the US state's weekly mortality predictions. Our model outperforms the existing baselines and fairly competes with the models adopted by CDC.

Future work can explore various association techniques for COVID-19 forecasting and extend the study to other diseases with complex spatiotemporal dynamics. Further research can focus on separate state-specific models, considering inherent differences and inter-state impacts. Overall, this work has the potential to provide accurate COVID-19 forecasting for spatiotemporal data, aiding in implementing effective measures to control the virus's spread.

References

1. CDC Covid Tracker: https://covid.cdc.gov/covid-data-tracker/#forecasting
2. Chimmula, V.K., Reddy, L.Z.: Time series forecasting of COVID-19 transmission in Canada using LSTM networks. Chaos Solitons Fractals **135**, 109864 (2020)
3. Kumar, S., et al.: Forecasting the spread of COVID-19 using LSTM network. BMC Bioinformatics **22**(6), 1–9 (2021)

4. Pavan, K., et al.: Forecasting the dynamics of COVID-19 pandemic in top 15 countries in April 2020: ARIMA model with machine learning approach. MedRxiv (2020)
5. Chinazzi, M., et al.: The effect of travel restrictions on the spread of the 2019 novel coronavirus (COVID-19) outbreak. Science **368**(6489), 395–400 (2020)
6. Ferguson, N.M., et al.: Strategies for mitigating an influenza pandemic. Nature **442**(7101), 448–452 (2006)
7. Kapoor, A., et al.: Examining covid-19 forecasting using spatio-temporal graph neural networks. arXiv preprint arXiv:2007.03113 (2020)
8. Mahmud, S., et al.: A human mobility data driven hybrid GNN+ RNN based model for epidemic prediction. In: 2021 IEEE International Conference on Big Data (Big Data). IEEE (2021)
9. Panagopoulos, G., Nikolentzos, G., Vazirgiannis, M.: Transfer graph neural networks for pandemic forecasting. In: Proceedings of the AAAI Conference on Artificial Intelligence, vol. 35, no. 6. (2021)
10. Rozemberczki, B., et al.: Pytorch geometric temporal: spatiotemporal signal processing with neural machine learning models. In: Proceedings of the 30th ACM International Conference on Information & Knowledge Management (2021)
11. Wang, Lijing, et al.: Using mobility data to understand and forecast COVID19 dynamics. medRxiv (2020)
12. Xue, J., et al.: Multiwave covid-19 prediction from social awareness using web search and mobility data. In: Proceedings of the 28th ACM SIGKDD Conference on Knowledge Discovery and Data Mining (2022)
13. Cramer, E.Y., et al.: Evaluation of individual and ensemble probabilistic forecasts of COVID-19 mortality in the United States. Proc. Natl. Acad. Sci. **119**(15), e2113561119 (2022)
14. Fritz, C., Dorigatti, E., Rögamer, D.: Combining graph neural networks and spatio-temporal disease models to predict covid-19 cases in Germany. arXiv preprint arXiv:2101.00661 (2021)
15. Cao, D., et al.: Spectral temporal graph neural network for multivariate time-series forecasting. Adv. Neural. Inf. Process. Syst. **33**, 17766–17778 (2020)
16. Bai, J., et al.: A3t-GCN: attention temporal graph convolutional network for traffic forecasting. ISPRS Int. J. Geo Inf. **10**(7), 485 (2021)
17. Kipf, T.N., Welling, M.: Semi-supervised classification with graph convolutional networks. arXiv preprint arXiv:1609.02907 (2016)
18. Vaswani, A., et al.: Attention is all you need. In: Advances in neural Information Processing Systems, vol. 30 (2017)
19. Learned-Miller, E.G.: Entropy and mutual information, p. 4. University of Massachusetts, Amherst, Department of Computer Science (2013)
20. Dong, E., Hongru, D., Gardner, L.: An interactive web-based dashboard to track COVID-19 in real time. Lancet. Infect. Dis **20**(5), 533–534 (2020)
21. Paszke, A., et al.: Pytorch: an imperative style, high-performance deep learning library. In: Advances in Neural Information Processing Systems, vol. 32 (2019)
22. Smyl, S., Ranganathan, J., Pasqua, A.: M4 forecasting competition: introducing a new hybrid ES-RNN model (2018). https://eng.uber.com/m4-forecasting-competition
23. Chicco, D., Warrens, M.J., Jurman, G.: The coefficient of determination R-squared is more informative than SMAPE, MAE, MAPE, MSE and RMSE in regression analysis evaluation. PeerJ Comput. Sci. **7**, e623 (2021)
24. JHU-IDD by Johns Hopkins University: Infectious Disease Dynamic Lab. https://github.com/HopkinsIDD/COVIDScenarioPipeline/

25. Computational Uncertainty Lab by Prof. Thomas McAndrew. https://zoltardata.com/model/737
26. Microsoft by Microsoft AI. https://www.microsoft.com/en-us/ai/ai-for-health/
27. Srivastava, A.: The Variations of SIkJalpha Model for COVID-19 Forecasting and Scenario Projections. arXiv preprint arXiv:2207.02919 (2022)
28. ESG by Robert Walraven. https://rwalraven.com/COVID19/
29. UCSD-NEU University of California, San Diego and Northeastern University. https://sites.google.com/view/yianma/epidemiology/
30. JHU-APL by Johns Hopkins University, Applied Physics Lab. https://buckymodel.com/
31. Karlen by Karlen Working Group. https://pypm.github.io/home/
32. MOBS by Northeastern University, Laboratory for the Modeling of Biological and Socio-technical Systems. https://covid19.gleamproject.org/
33. GT-DeepCOVID by Georgia Institute of Technology, College of Computing. https://deepcovid.github.io/
34. BPagano by Bob Pagano. https://bobpagano.com/
35. Ray, E.L., et al.: Ensemble forecasts of coronavirus disease 2019 (COVID-19) in the US. MedRXiv (2020)
36. Columbia by Columbia University. https://columbia.maps.arcgis.com/apps/webappviewer/index.html?id=ade6ba85450c4325a12a5b9c09ba796c

IoT Lakehouse: A New Data Management Paradigm for AIoT

Guochuan Liu[1], Zhenjiang Pang[1,2], Jing Zeng[1(✉)], Haimin Hong[1], Yongming Sun[1], Mingjie Su[1], and Nan Ma[1]

[1] China Gridcom Co., Ltd., Shenzhen, China
jerryzengjing@163.com
[2] Shenzhen SmartChip Microelectronics Technology Co., Ltd., Shenzhen, China

Abstract. The Internet of Things (IoT) and Artificial Intelligence of Things (AIoT) are emerging as promising paradigms for enabling ubiquitous and intelligent applications across various domains. However, managing and utilizing the massive and heterogeneous data generated by IoT and AIoT devices poses significant challenges for traditional data management systems. In this paper, we present a new data management paradigm called IoT Lakehouse, which aims to integrate the best practices of data warehouse and data lake to provide a unified, scalable and efficient platform for IoT and AIoT data. We define the concept and characteristics of IoT Lakehouse, and compare it with other existing data management paradigms. We present a refercence architecture and key technologies of IoT Lakehouse, and discuss how it supports various needs and scenarios of AIoT. We also analyze the main challenges and future directions of IoT Lakehouse research and development.

Keywords: Lakehouse · AIoT · Data Platform

1 Introduction

The Internet of Things (IoT) is a network of interconnected devices that can sense, communicate and act on the physical and cyber world. IoT has been widely applied in various domains, such as smart city, smart home, smart health, smart agriculture, smart manufacturing, etc., to provide innovative and convenient services for human beings. According to a report by Statista [7], the number of IoT devices worldwide is expected to reach 75.44 billion by 2025, generating a huge amount of data.

However, the data generated by IoT devices are often massive, heterogeneous, complex and dynamic, which pose significant challenges for traditional data management systems. For example, how to efficiently store and access the IoT data with different formats, schemas and quality? How to process and analyze the IoT data with different latency, granularity and complexity requirements? How to ensure the security and privacy of the IoT data in a distributed and open environment?

S. Zhang et al. (Eds.): BigData 2023, LNCS 14203, pp. 34–47, 2023.
https://doi.org/10.1007/978-3-031-44725-9_3

To address these challenges, a new paradigm called Artificial Intelligence of Things (AIoT) has emerged, which aims to integrate artificial intelligence (AI) techniques with IoT devices to enable intelligent and autonomous applications. AIoT can enhance the capabilities of IoT devices in terms of perception, cognition, decision making and action. AIoT can also provide insights and value from the IoT data through various methods, such as machine learning, deep learning, computer vision, natural language processing, etc. According to a report by MarketsandMarkets [1], the global AIoT market size is projected to grow from USD 5.1 billion in 2019 to USD 16.2 billion by 2024, at a compound annual growth rate (CAGR) of 26.0%. However, managing and utilizing the data for AIoT applications also poses new challenges for traditional data management systems. For example, how to support the data lifecycle of AIoT applications from data collection to data consumption? How to enable the collaboration and interoperability of AIoT devices and applications? How to balance the trade-offs between edge computing and cloud computing for AIoT data processing and analysis?

To counter these issues, a new data management paradigm called Lakehouse [20] can offers a best solution, which aims to integrate the best practices of data warehouse and data lake to provide a unified, scalable and efficient platform for IoT and AIoT data. Data warehouse is a centralized repository that stores structured and curated data for business intelligence and analytics. Data lake is a distributed repository that stores raw and diverse data for exploratory analysis and machine learning. IoT Lakehouse combines the advantages of both paradigms to support both structured and unstructured data, both batch and stream processing, both schema-on-write and schema-on-read, both descriptive and predictive analytics.

In this paper, we discuss the concept, characteristics of Lakehouse for IoT. We also compare Lakehouse with other existing data management paradigms in IoT. Furthermore we propose a reference architecture for IoT Lakehouse and discuss how Lakehouse supports various needs and scenarios of AIoT applications. We also analyze the main challenges and future directions of IoT Lakehouse research and development.

The rest of this paper is organized as follows: Sect. 2 introduces the concept and characteristics of IoT Lakehouse, and then compares IoT Lakehouse with other data management paradigms. Section 3 introduces the proposed architecture and enabling technologies of IoT Lakehouse. Sectionn 4 discusses how IoT Lakehouse supports various needs and scenarios of AIoT applications. Sectionn 5 analyzes the main challenges and future directions of IoT Lakehouse research and development. Section 6 concludes this paper.

2 The IoT Lakehouse

2.1 Concept and Characteristics of IoT Lakehouse

IoT Lakehouse is a new data management paradigm that aims to integrate the best practices of data warehouse and data lake to provide a unified, scalable and efficient platform for IoT and AIoT data. An IoT Lakehouse is a data management system that supports both structured and unstructured data, both batch and stream processing, both schema-on-write and schema-on-read, both descriptive and predictive analytics for IoT and AIoT applications.

The main characteristics of IoT Lakehouse are:

- Unified: IoT Lakehouse provides a single platform that can store and manage both structured and unstructured data from various sources and formats, such as sensors, cameras, RFID tags, smartphones, etc. IoT Lakehouse also provides a unified interface that can support both SQL and NoSQL queries, as well as various programming languages and frameworks for data processing and analysis.
- Scalable: IoT Lakehouse can scale horizontally and vertically to handle the massive and dynamic data generated by IoT and AIoT devices. IoT Lakehouse can leverage distributed computing technologies such as Hadoop, Spark, Kafka, etc., to enable parallel and distributed data processing and analysis. IoT Lakehouse can also leverage cloud computing technologies such as AWS, Azure, Google Cloud, etc., to provide elastic and on-demand resources for data storage and computation.
- Efficient: IoT Lakehouse can optimize the performance and cost of data management for IoT and AIoT applications. IoT Lakehouse can use various techniques such as compression, partitioning, indexing, caching, etc., to reduce the storage space and query latency of data. IoT Lakehouse can also use various techniques such as materialized views, incremental updates, delta lake, etc., to ensure the consistency and quality of data. IoT Lakehouse can also use various techniques such as query optimization, query federation, query rewriting, etc., to improve the efficiency and accuracy of data analysis.
- Flexible: IoT Lakehouse can support both schema-on-write and schema-on-read for data management. Schema-on-write means that the data is validated and transformed according to a predefined schema before being stored in the system. Schema-on-read means that the data is stored in its raw form without any schema validation or transformation, and the schema is applied only when the data is read from the system. Schema-on-write can ensure the quality and consistency of data, while schema-on-read can enable the exploration and discovery of data.
- Versatile: IoT Lakehouse can support both descriptive and predictive analytics for IoT and AIoT applications. Descriptive analytics means that the system can provide reports and dashboards that summarize the past or current state of the data. Predictive analytics means that the system can provide models and algorithms that predict the future or unknown state of the data. Descriptive analytics can help users understand what happened or what is happening in the data, while predictive analytics can help users understand why it happened or what will happen in the data.

2.2 Comparison with Other Data Management Paradigms

IoT Lakehouse is not the first nor the only data management paradigm for IoT and AIoT applications. There are other existing paradigms [15] that have been proposed or adopted in practice, such as data warehouse, data lake, data mesh, etc. In this subsection, we compare IoT Lakehouse with these paradigms in terms of their definitions, advantages and disadvantages, as Table 1 shown.

Table 1. The Comparison with Other Data Management Paradigms

Data Management Paradigm	Description	Advantages	Disadvantages
Data warehouse	Centralized repository	The quality and consistency of data	Costly and inflexible
Data mesh	Decentralized and distributed architecture	Autonomy and agility of data producers and consumers	Challenges of data governance, discovery, integration and quality across different domains or teams
IoT lakehouse	Hybrid paradigm support structured and unstructured data	Both batch and stream processing	Complex and challenging to design and implement

Data Warehouse [10]: A data warehouse is a centralized repository that stores structured and curated data for business intelligence and analytics.A data warehouse follows a schema-on-write approach, which means that the data is validated and transformed according to a predefined schema before being stored in the system. A data warehouse also follows a batch processing approach, which means that the data is processed periodically in large batches. Its advantages is that a data warehouse can ensure the quality and consistency of data, as well as provide fast and reliable queries for descriptive analytics. The disadvantages are that a data warehouse can be costly and complex to build and maintain, as well as inflexible and rigid to accommodate new or changing data sources or formats. A data warehouse also cannot support unstructured or semi-structured data, nor stream processing or predictive analytics.

Data Mesh [13]: A data mesh is a decentralized and distributed architecture that treats data as a product that can be owned and managed by different domains or teams. A data mesh follows a schema-on-write approach for each data product, which means that the data is validated and transformed according to a domain-specific schema before being stored in the system. A data mesh also follows a stream processing approach, which means that the data is processed continuously and incrementally. Its advantages are that a data mesh can enable the autonomy and agility of data producers and consumers, as well as provide real-time and contextualized data for descriptive and predictive analytics. And the disadvantages are that a data mesh can introduce the challenges of data governance, discovery, integration and quality across different domains or teams. A data mesh also requires a high level of collaboration and coordination among data producers and consumers.

IoT Lakehouse: An IoT Lakehouse is a hybrid paradigm that combines the best practices of data warehouse and data lake to provide a unified, scalable and efficient platform for IoT and AIoT data. An IoT Lakehouse follows both schema-on-write and schema-on-read approaches, which means that the data can be stored in both structured and unstructured forms, and the schema can be applied either before or after the storage. An IoT Lakehouse also follows both batch and stream processing approaches, which means that the data can be processed both periodically and continuously. Its advantages are that an IoT Lakehouse can support both structured and unstructured data, both batch and stream processing, both schema-on-write and schema-on-read, both descriptive and predictive analytics for IoT and AIoT applications. The disadvantages are that an IoT Lakehouse can be complex and challenging to design and implement, as well as requires a balance between the trade-offs of different approaches. An IoT Lakehouse also needs to address the issues of data quality, security, privacy, scalability, interpretability, etc.

3 A Reference Architecture and Enabling Technologies for IoT Lakehouse

IoT Lakehouse is a new data management paradigm for AIoT that combines the benefits of data lakes and data warehouses. It enables scalable, efficient, and unified analytics on diverse and massive IoT data sources. In this chapter, we present a reference architecture and technologies of IoT lakehouse, based on the following principles:

Data is stored in its native format in a central data lake that supports various data types and structures, such as sensor data, video data, text data, etc. Data is processed and transformed using various computing frameworks, such as batch processing, stream processing, and ad hoc query, to support different analytics needs and latency requirements. Data is exposed and consumed through various service layers that provide SQL APIs and declarative dataframe APIs for easy access and integration with applications and tools. Data is governed and secured

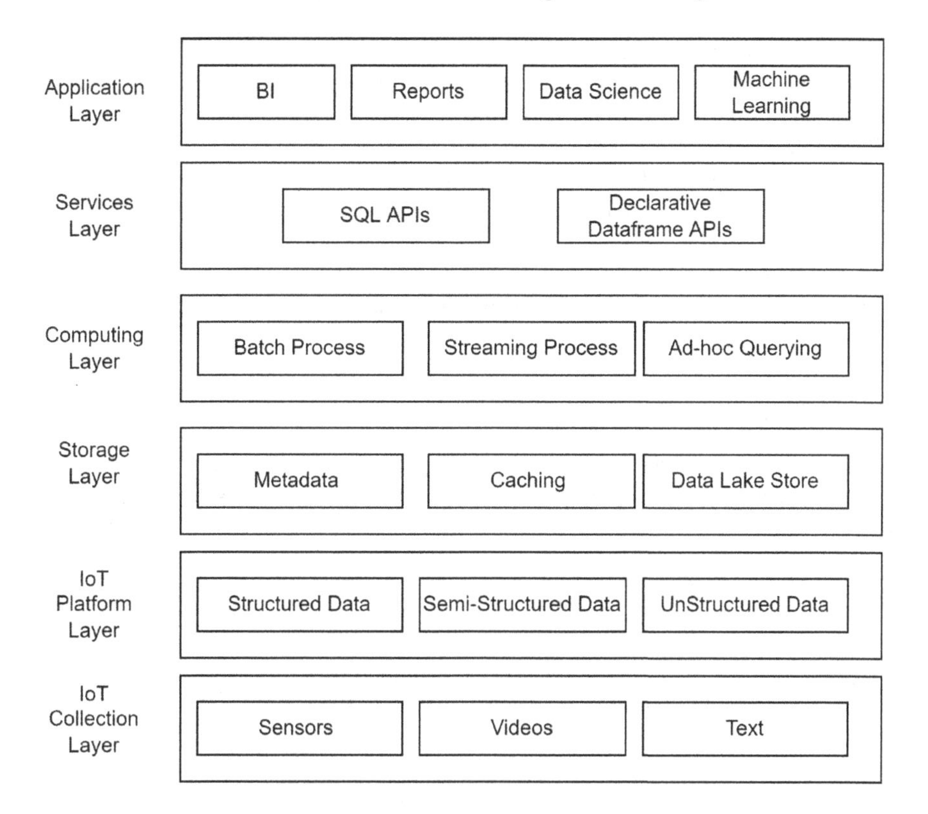

Fig. 1. A Reference Architecture for IoT Lakehouse.

using metadata management, caching management, access control, encryption, etc.

As the Fig. 1 shown, the lakehouse architecture consists of six layers from bottom to top: IoT collection layer, IoT platform layer, storage layer, computing layer, service layer, and application layer. The following table describes the components and functions of each layer:

3.1 IoT Collection Layer

This layer collects data from various IoT sources using different data formats and protocols. Some of the technologies that are used in this layer are:

- Sensors: These are devices that measure physical quantities, such as temperature, humidity, pressure, etc., and convert them into electrical signals. Sensors can be wired or wireless, passive or active, analog or digital, etc.
- Cameras: These are devices that capture images or videos of the surrounding environment. Cameras can have different resolutions, frame rates, lenses, etc.
- Text Sources: These are sources that generate text data, such as documents, emails, social media posts, etc. Text data can have different languages, formats, styles, etc.

- Data Compression: This is a technique that reduces the size of data by removing redundant or irrelevant information. Data compression can improve the efficiency and speed of data transmission and storage.
- Data Encryption: This is a technique that protects the confidentiality and integrity of data by transforming it into an unreadable form using a secret key. Data encryption can prevent unauthorized access and modification of data.
- Data Validation: This is a technique that checks the quality and accuracy of data by applying rules or criteria. Data validation can detect and correct errors or anomalies in data.

3.2 IoT Platform Layer

This layer connects and manages the IoT devices and their data using various software platforms and services [17]. Some of the technologies that are used in this layer are:

- Device Management: This is a service that handles the registration, authentication, configuration, monitoring, and control of IoT devices. It can also perform firmware updates, device diagnostics, device grouping, etc.
- Protocol Adaptation: This is a service that adapts different communication protocols used by IoT devices, such as MQTT, CoAP, HTTP, etc. It can also perform protocol conversion, message routing, message filtering, etc.
- Rule Management: This is a service that defines and executes rules or actions based on IoT data or events. It can also perform event processing, complex event processing (CEP), event-driven architecture (EDA), etc.
- IoT Platforms: These are software platforms that provide end-to-end solutions for connecting, managing, and analyzing IoT data. They typically include features such as device management, protocol adaptation, rule management, data ingestion, data processing, data visualization, and data integration. Some of the popular IoT platforms are Azure IoT Hub, AWS IoT Core, Google Cloud IoT Core, IBM Watson IoT Platform, etc.

3.3 Storage Layer

This layer stores all kinds of IoT data in its native format in a central data lake using various storage technologies and services. Some of the technologies that are used in this layer are:

- Metadata Management: This is a service that manages the metadata of the data lake, such as schema, lineage, quality, etc. It can also provide cataloging, indexing, searching, and discovery features. Some of the technologies that are used for metadata management for lakehouse are:1) Delta Lake [6]: This is an open source project that provides a metadata layer for data lakes. Delta Lake tracks which files are part of different table versions and offers rich management features like ACID-compliant transactions, time travel, schema enforcement and evolution, and data validation. 2)Unity Catalog [5]: This is a unified

governance solution for lakehouse that provides a unified data access layer and a centralized mechanism for managing data governance and access controls. Unity Catalog supports three-tier namespacing (catalog.database.table) for organizing and granting access to data. It also supports external locations and storage credentials as securable objects. 3)Apache Atlas [14]: This is an open source project that provides scalable governance for data lakes. Apache Atlas enables metadata management and governance across different data platforms and processes. It also provides lineage tracking, impact analysis, data classification, and security integration.
- Caching Management: This is a service that manages the caching of frequently accessed or hot data for faster access and lower latency. It can also provide caching policies, eviction strategies, and consistency guarantees. The typical solution to lakehouse is Alluxio [16] which is widely used to caching for improving the efficiency to data lake files and enabling the separations of storage and computing.
- Data Lake Storage: This is a storage technology that stores the raw or processed IoT data in its native format using a distributed file system or object storage. It can also provide encryption, access control, auditing, and firewall features. Some of the popular data lake storage technologies are Azure Data Lake Storage (ADLS), Amazon Simple Storage Service (S3), Hadoop Distributed File System (HDFS), etc. The major management framworks for data lake are Delta [2], Iceberg [4] and Hudi [3], which are contructed at the upper of these data lake storage and provide the management of table format.

3.4 Computing Layer

This layer transforms and analyzes IoT data stored in the data lake using various computing frameworks and tools. Some of the technologies that are used in this layer are:

- Batch Processing [19]: This is a computing framework that processes large volumes of IoT data in batches using frameworks such as Spark or Hadoop. It can also perform data quality, data cleansing, and data enrichment tasks.
- Stream Processing [8]: This is a computing framework that processes real-time or near-real-time IoT data streams using frameworks such as Flink or Kafka Streams. It can also perform event processing, complex event processing (CEP), event-driven architecture (EDA), etc.
- Ad hoc Query [18]: This is a computing framework that executes interactive queries on IoT data using frameworks such as Presto or Hive. It can also perform SQL queries, OLAP queries, BI queries, etc.

3.5 Service Layer

This layer exposes and consumes IoT data from the data lake using various APIs and interfaces. Some of the technologies [20] that are used in this layer are:

- SQL APIs: These are APIs that provide SQL interfaces for accessing and analyzing IoT data using standard or extended SQL syntax. They can also provide JDBC/ODBC drivers, RESTful APIs, GraphQL APIs, etc.
- Declarative Dataframe APIs: These are APIs that provide declarative interfaces for accessing and analyzing IoT data using dataframe abstractions and operations. They can also provide Python APIs, R APIs, Scala APIs, Java APIs, etc.

3.6 Application Layer

This layer provides various functionalities and applications for the end users and customers to interact with the IoT data and analytics. Some of the technologies that are used in this layer are:

- Business Intelligence (BI): This is a technology that provides tools and methods for analyzing, visualizing, and reporting IoT data. It can also provide dashboards, charts, graphs, tables, etc.
- Reporting: This is a technology that provides tools and methods for generating and delivering reports based on IoT data. It can also provide templates, formats, schedules, etc.
- Data Science: This is a technology that provides tools and methods for applying scientific methods and techniques to IoT data. It can also provide statistics, mathematics, machine learning, etc.
- Machine Learning (ML): This is a technology that provides tools and methods for creating and applying models that can learn from IoT data and make predictions or decisions. It can also provide supervised learning, unsupervised learning, reinforcement learning, etc.

4 The Applications of IoT Lakehouse

4.1 Industrial IoT

Industrial IoT (IIoT) [12] refers to the use of IoT technologies in industrial sectors, such as manufacturing, logistics, energy, healthcare, etc. IIoT can enable various use cases, such as predictive maintenance, quality control, remote monitoring, asset optimization, fleet management, etc. IIoT can also generate large volumes and varieties of data from sensors, cameras, machines, devices, etc.

IoT lakehouse can provide a powerful solution for building and scaling IIoT applications on a single platform. It can store and process various types of data in its native format, such as sensor data, video data, text data, etc. It can also provide batch processing, stream processing, and ad hoc query capabilities for different analytics needs and latency requirements. It can also expose and consume data through SQL APIs and declarative dataframe APIs for easy access and integration with applications and tools.

For example, a manufacturing company can use IoT lakehouse to collect and analyze data from its shop floor machines and devices. It can use batch processing to perform data quality, data cleansing, and data enrichment tasks on the raw data. It can use stream processing to perform real-time event processing and complex event processing on the data streams. It can use ad hoc query to execute interactive queries on the data using SQL or dataframe syntax. And it can also use SQL APIs and declarative dataframe APIs to access and analyze the data using BI tools or ML models. By using IoT lakehouse, the company can improve its operational efficiency, product quality, and customer satisfaction.

4.2 Smart City

Smart city [9] refers to the use of IoT technologies to enhance the quality and performance of urban services, such as transportation, energy, water, waste management, public safety, etc. Smart city can enable various use cases, such as traffic management, smart parking, smart lighting, smart metering, smart waste management, smart surveillance. Smart city can also generate large volumes and varieties of data from sensors, cameras, vehicles, devices, etc.

IoT lakehouse can provide a powerful solution for building and scaling smart city applications on a single platform. It can store and process various types of data in its native format, such as sensor data, video data, text data, etc. It can also provide batch processing, stream processing, and ad hoc query capabilities for different analytics needs and latency requirements of smart city. It can also expose and consume data through SQL APIs and declarative dataframe APIs for easy access and integration with applications and tools.

For example, a city government can use IoT lakehouse to collect and analyze data from its traffic cameras and sensors. It can use batch processing to perform data quality, data cleansing, and data enrichment tasks on the raw data. It can use stream processing to perform real-time event processing and complex event processing on the data streams of smart city applications. It can use ad hoc query to execute interactive queries on the data using SQL or dataframe syntax. It can also use SQL APIs and declarative dataframe APIs to access and analyze the data using BI tools or ML models. By using IoT lakehouse, the city government can improve its traffic efficiency, safety, and sustainability.

4.3 Healthcare IoT

Healthcare IoT [9] refers to the use of IoT technologies to improve the quality and efficiency of healthcare services, such as diagnosis, treatment, monitoring, prevention, etc. Healthcare IoT can enable various use cases, such as remote patient monitoring, telemedicine, wearable devices, smart pills, smart implants, etc. Healthcare IoT can also generate large volumes and varieties of data from sensors, cameras, devices, medical records, etc.

IoT lakehouse can provide a powerful solution for building and scaling healthcare IoT applications on a single platform. It can store and process various types of data in its native format, such as wearable sensor data, video data, medical text data, etc. It can also provide batch processing, stream processing, and ad hoc query capabilities for different analytics needs latency requirements of heathcare IoT data. It can also expose and consume data through SQL APIs and declarative dataframe APIs for easy access and integration with medical applications.

For instance, a healthcare provider can use IoT lakehouse to collect and analyze data from its wearable devices and medical records. It can use batch processing to perform data quality, data cleansing, and data enrichment tasks on the raw medical data. It can use stream processing to perform real-time event processing and complex event processing on the data streams. It can use ad hoc query to execute interactive queries on the data using SQL for heatlh application developers. By using IoT lakehouse, the healthcare provider can greatly enhance its patient care, outcomes, and satisfaction.

5 Chanllenges and Future Directions of IoT Lakehouse

IoT Lakehouse is a new data management paradigm that combines the advantages of data lakes and data warehouses for AI-enabled IoT applications. IoT Lakehouse allows analytics on the most complete and up-to-date data from various sources, such as sensors, devices, applications, and enterprise systems. It also supports collaborative data science and machine learning on large-scale data using open source frameworks and libraries. However, IoT Lakehouse also faces several challenges and opportunities for future research and development. Some of these challenges are:

- Data quality and governance: How to ensure the reliability, consistency, security, and privacy of data across different sources and formats in IoT Lakehouse? How to implement effective data governance policies and mechanisms for data access, sharing, and usage in IoT Lakehouse? a typical solution is to use metadata management, data cataloging, data lineage, data quality assessment, data masking, encryption, anonymization, access control, auditing, and compliance tools to ensure data quality and governance in IoT Lakehouse.
- Data integration and transformation: How to efficiently ingest, process, transform, and store data from heterogeneous and dynamic IoT sources in IoT Lakehouse? How to handle the complexity, diversity, velocity, and volume of IoT data in IoT Lakehouse? For these challenges, we can use scalable, distributed, parallel, streaming, batch, or hybrid data processing frameworks such as Apache Spark or Apache Flink to integrate and transform data from various IoT sources in IoT Lakehouse, and use schema inference, schema evolution, schema validation, schema registry, or schema-on-read techniques to handle diverse and evolving data formats in IoT Lakehouse.
- Data analysis and AI: How to leverage the latest advances in data analysis and AI techniques for IoT applications in IoT Lakehouse? How to optimize the performance, scalability, and cost of data analysis and AI workloads in IoT Lakehouse? How to enable interoperability and compatibility of different data analysis and AI frameworks and tools in IoT Lakehouse? Large language model, such as GPT 4 [11], can be imported to help assist to perform data analysis and AI tasks on large-scale data in IoT Lakehouse. Also we

can use auto-scaling, auto-tuning, auto-termination, or serverless technologies to optimize the performance, scalability, and cost of data analysis and AI workloads in IoT Lakehouse, and leverage common APIs or interfaces such as Apache Arrow or MLflow to enable interoperability and compatibility of different data analysis and AI frameworks and tools in IoT Lakehouse.
- Data value and monetization: How to measure and maximize the value of data in IoT Lakehouse? How to create new business models and revenue streams based on data in IoT Lakehouse? How to balance the trade-offs between data value and data cost in IoT Lakehouse? Using data valuation methods such as market-based approach or income-based approach to measure the value of data in IoT Lakehouse. A candidate solution is to use data monetization strategies such as selling or licensing data products or services based on data in IoT Lakehouse, and use cost optimization techniques such as tiered storage or compression to balance the trade-offs between data value and data cost in IoT Lakehouse.

Some of the possible future directions for IoT Lakehouse are:

- Developing new architectures, algorithms, methods, and systems for addressing the challenges and opportunities in IoT Lakehouse.
- Exploring new use cases and domains for applying IoT Lakehouse such as industrial IoT (IIoT), smart cities (SC), healthcare (HC), transportation (TR), energy (EN), etc.
- Benchmarking and evaluating the performance, effectiveness, efficiency, and impact of IoT Lakehouse solutions for different scenarios and requirements.
- Establishing standards, best practices, guidelines, and frameworks for designing implementing operating and managing IoT Lakehouse solutions.

6 Conclusions

In this paper, we have surveyed the IoT Lakehouse, a new data management paradigm that combines the advantages of data lakes and data warehouses for AI-enabled IoT applications. We have discussed the main characteristics, benefits, and challenges of IoT Lakehouse, as well as some of the existing solutions and frameworks that support it. We have also presented a reference architecture for IoT Lakehouse, and identified some of the future directions and opportunities for research and development in IoT Lakehouse. We believe that IoT Lakehouse is a promising and emerging paradigm that can enable scalable, reliable, and efficient data analysis and AI for various IoT use cases and domains. We hope that this paper can provide a comprehensive overview and reference for researchers, practitioners, and students who are interested in IoT Lakehouse.

References

1. Artificial Intelligence in IoT Market by Component. https://www.marketdataforecast.com/market-reports/artificial-intelligence-in-iot-market. Accessed 7 May 2023

2. Delta: Build Lakehouses with Delta Lake. https://delta.io/. Accessed 7 May 2023
3. Hudia:a transactional data lake platform. https://hudi.apache.org/. Accessed 7 May 2023
4. Iceberg: The open table format for analytic datasets. https://iceberg.apache.org/. Accessed 7 May 2023
5. What is a Data Lakehouse? - Databricks. https://www.databricks.com/glossary/data-lakehouse. Accessed 8 May 2023
6. What is the Databricks Lakehouse? - Azure Databricks. https://learn.microsoft.com/en-us/azure/databricks/lakehouse/. Accessed 8 May 2023
7. Number of IoT connected devices worldwide 2019–2030. https://www.statista.com/statistics/1183457/iot-connected-devices-worldwide/. Accessed 7 May 2023
8. Akidau, T., et al.: The dataflow model: a practical approach to balancing correctness, latency, and cost in massive-scale, unbounded, out-of-order data processing. Proc. VLDB Endow. **8**(12), 1792–1803 (2015). https://doi.org/10.14778/2824032.2824076
9. Asad, U., Mohammed, A.S.: Deep learning and industrial internet of things to improve smart city safety. In: 2023 International Conference on Business Analytics for Technology and Security (ICBATS), pp. 1–10 (2023). https://doi.org/10.1109/ICBATS57792.2023.10111164
10. Azevedo, R., Silva, J.P., Lopes, N., Curado, A., Nunes, L.J., Lopes, S.I.: Designing an IoT-enabled data warehouse for indoor radon time series analytics. In: 2022 17th Iberian Conference on Information Systems and Technologies (CISTI), pp. 1–6 (2022). https://doi.org/10.23919/CISTI54924.2022.9820540
11. Bubeck, S., et al.: Sparks of artificial general intelligence: Early experiments with gpt-4 (2023), https://www.microsoft.com/en-us/research/publication/sparks-of-artificial-general-intelligence-early-experiments-with-gpt-4/
12. Cunha, B., Sousa, C.: On the definition of intelligible IIoT architectures. In: 2021 16th Iberian Conference on Information Systems and Technologies (CISTI), pp. 1–6 (2021). https://doi.org/10.23919/CISTI52073.2021.9476342
13. Dehghani, Z.: Data Mesh. O'Reilly Media (2022). https://books.google.de/books?id=jmZjEAAAQBAJ
14. Gallas, E.J., Malon, D., Hawkings, R.J., Albrand, S., Torrence, E.: An integrated overview of metadata in atlas (2010)
15. Harby, A.A., Zulkernine, F.: From data warehouse to lakehouse: A comparative review. In: 2022 IEEE International Conference on Big Data (Big Data), pp. 389–395 (2022). https://doi.org/10.1109/BigData55660.2022.10020719
16. Li, H.: Alluxio: A Virtual Distributed File System. Ph.D. thesis, EECS Department, University of California, Berkeley (2018). https://www2.eecs.berkeley.edu/Pubs/TechRpts/2018/EECS-2018-29.html
17. Macagnano, D., Destino, G., Abreu, G.: Indoor positioning: a key enabling technology for IoT applications. In: 2014 IEEE World Forum on Internet of Things (WF-IoT), pp. 117–118 (2014). https://doi.org/10.1109/WF-IoT.2014.6803131
18. Sethi, R., et al.: Presto: SQL on everything. In: 2019 IEEE 35th International Conference on Data Engineering (ICDE), pp. 1802–1813 (2019). https://doi.org/10.1109/ICDE.2019.00196
19. Zaharia, M., et al.: Resilient distributed datasets: a fault-tolerant abstraction for in-memory cluster computing. In: Proceedings of the 9th USENIX Conference on Networked Systems Design and Implementation. p. 2. NSDI'12, USENIX Association, USA (2012)

20. Zaharia, M., Ghodsi, A., Xin, R., Armbrust, M.: Lakehouse: a new generation of open platforms that unify data warehousing and advanced analytics. In: 11th Conference on Innovative Data Systems Research, CIDR 2021, Virtual Event, 11–15 January 2021, Online Proceedings. https://www.cidrdb.org (2021), https://cidrdb.org/cidr2021/papers/cidr2021_paper17.pdf

HMGR: A Hybrid Model for Geolocation Recommendation

Ruwang Wen[1], Zhengxiang Cheng[1], Weixuan Mao[1], Zhuolin Mei[1,2,3](✉), Jiaoli Shi[1,2,3], and Xiao Cheng[1]

[1] School of Computer and Big Data Science, Jiujiang University, Jiujiang, Jiangxi, China
meizhuolin@126.com
[2] Jiujiang Key Laboratory of Network and Information Security, Jiujiang, Jiangxi, China
[3] Institute of Information Security, Jiujiang University, Jiujiang, Jiangxi, China

Abstract. The Geosocial networks integrate geographical location information into traditional social networks, bridging the gap between people's real-life experiences and the virtual world. As a significant application of geosocial networks, location recommendation suggests places that individuals may find interesting, offering valuable references for their travels and greatly enhancing their lives. Consequently, the challenge of recommending relevant locations to users from a vast pool of geographic options has become a prominent topic in academic research. Collaborative filtering algorithms stand as one of the classic solutions in the field of recommendations. While they partially address the problem of information overload, they often encounter a common obstacle known as the cold start problem. To overcome this issue, this study makes two primary contributions: Firstly, it proposes the utilization of Markov chains to mitigate the cold start problem. Secondly, it introduces a hybrid recommendation model called HMGR for location-based recommendations, which effectively enhances the accuracy of suggestions. We evaluate the efficacy of the Markov chain and HMGR model through extensive experimentation. The results demonstrate that the implementation of Markov chains successfully alleviates the cold start problem, and our HMGR model significantly improves the precision of recommendations.

Keywords: Geographical Location · Collaborative Filtering · Markov Chain · Cold Start

1 Introduction

In the context of information overload [1, 2], users are faced with an overwhelming amount of information and options, making it difficult to accurately find the desired information or make the best choices. In such a situation, both probabilistic relational database hypothesis querying [3] and recommendation systems can help users more effectively obtain useful information and address uncertainty. However, recommendation systems have clear advantages in dealing with information overload. They have the capability to provide personalized recommendations based on users' historical behavior and interests. This means that users will receive content that matches their preferences

S. Zhang et al. (Eds.): BigData 2023, LNCS 14203, pp. 48–62, 2023.
https://doi.org/10.1007/978-3-031-44725-9_4

and needs, greatly enhancing the efficiency and satisfaction of information retrieval. Recommendation systems can leverage image processing techniques [4, 5] to extract content features from images that users post on social media. Furthermore, they can fully utilize users' historical data for personalized recommendations. By analyzing users' behavior data, such as click history, purchase records, search history, comments, and favorites, recommendation systems can gain deep insights into users' preferences, habits, and needs, further increasing user satisfaction and click-through rates.

However, recommendation systems face a significant challenge in practical applications, known as the cold-start problem [6]. The cold-start problem refers to the lack of sufficient historical data for new users or new items, making it difficult for the recommendation system to accurately predict their preferences. In real-world applications, new users and items are inevitable and may occur frequently. If the recommendation system cannot effectively address the cold-start problem, these new users and items will not receive personalized recommendations, which can lower user experience and the utility of the recommendation system.

Additionally, recommendation accuracy is of utmost importance for the value of a recommendation system. Accurate recommendations not only increase user trust in the recommendation system but also significantly improve user satisfaction and click-through rates. On the contrary, poor recommendation accuracy may lead to user dissatisfaction and even prompt users to discontinue using the recommendation service, resulting in potential economic losses for businesses. Therefore, improving recommendation accuracy is a key goal in recommendation system research, especially in highly competitive market environments, where precise recommendations are essential for gaining a competitive advantage. While some new solutions have shown promising results in addressing the cold-start problem and improving accuracy in recent years, there are still limitations that need to be further addressed.

The core idea of content-based recommendation systems [7] is to utilize item content information for personalized recommendations, which builds upon the advancements in information filtering technology. By employing machine learning techniques, these systems can extract user preferences from the descriptive features of the items, thereby alleviating the cold-start problem without relying solely on user ratings. However, it is worth noting that many existing content-based recommendation methods [8–10] face challenges in uncovering users' latent interests and encounter difficulties in accurately extracting item features, which can ultimately lead to lower recommendation accuracy.

In order to uncover a user's potential interests, a collaborative filtering recommendation [11] was proposed. This method assumes that users with similar interests may have similar preferences toward similar items. The core idea is to use a neighbor-based recommendation algorithm that leverages similarity measures between users or items and historical behavior data [12]. However, traditional collaborative filtering recommendation systems are facing challenges due to the exponential growth of data from various applications and services, leading to issues like the cold-start problem [6]. To address the cold-start problem and enhance recommendations, researchers have explored utilizing user social relationships for recommendation [13–15]. Friend recommendations can increase trust and help alleviate the cold-start problem. However, the collaborative filtering algorithm is vulnerable to data bias, as it tends to favor popular items that receive more attention and evaluations. This bias can lead to recommendation results that are biased towards these popular items, overlooking individualized preferences.

To enhance recommendation efficiency, hybrid recommendation [16] approaches have been proposed [17–19]. The main objective of hybrid recommendation algorithms is to leverage the strengths of various recommendation algorithms while mitigating their limitations. By integrating different approaches, hybrid recommendations can effectively address the cold-start problem and provide diverse recommendations. However, it is important to acknowledge that current hybrid recommendation methods may encounter challenges such as increased complexity, longer recommendation times, and difficulties in achieving a balanced integration. Despite these challenges, hybrid recommendation techniques have demonstrated promising results in improving overall recommendation performance.

Considering the limitations of previous approaches, we propose a hybrid model for geolocation recommendation, namely HMGR. In our model, we first assess the user's historical data volume. If the user is identified as new, the system employs a Markov chain for personalized recommendations. Conversely, for existing users, collaborative filtering is utilized for recommendation generation. Through extensive simulation experiments, we have observed that HMGR effectively addresses the challenges associated with the cold-start problem and significantly improves the accuracy of recommendation results.

We summarize our proposed HMGR model based on the following three key aspects.

(1) We mitigate the cold start problem in geolocation-based recommendations by utilizing Markov chains.
(2) We propose a hybrid recommendation model called HMGR for geolocation-based recommendations, which combines the use of Markov chains and collaborative filtering.
(3) Conduct extensive experiments to evaluate the efficiency and effectiveness of the HMGR model, along with a thorough analysis.

The remaining sections of this paper are structured as follows: Sect. 2 provides an overview of related research. Section 3 introduces the essential preliminary knowledge. Section 4 describes the construction of our proposed model. In Sect. 5, we present the experimental evaluation. Finally, Sect. 6 summarizes the findings and conclusions of this study.

2 Related Work

Traditional recommendation algorithms can be classified into three main categories: content-based recommendation algorithms [7], collaborative filtering recommendation algorithms [11], and hybrid recommendation algorithms [16].

2.1 Content-Based Recommendation

Content-based recommendation, as the name implies, is a method that recommends items based on their similarity in terms of content. It can be considered one of the founding approaches in recommendation algorithms, as it relies on the premise that items with similar content to a user's previously preferred items are likely to be of interest [20]. The concept of content-based recommendation originated from research in information retrieval [7]. With the rapid advancement of information retrieval and the widespread adoption of applications like email, content-based recommendation has found extensive application in the field. Content-based recommendation primarily involves the description of item content features and user profiles (such as interests and preferences). It efficiently filters and selects more valuable information, offering several advantages such as high recommendation efficiency, not requiring user evaluations or additional information, and mitigating the cold-start problem associated with new items [8–10].

For example, literature [9] proposes a content-based recommendation system model is proposed. This model tackles the cold start problem by employing content-based methods, which involves making recommendations for new users or new items during the recommendation process. Moreover, the model takes into account the incorporation of features like security, reliability, and transparency in career recommendation, assisting students in making informed career choices. However, despite these advancements, these studies still encounter challenges, such as limited diversity in recommended content and complexities in extracting multiple item features.

2.2 Collaborative Filtering Recommendation

In order to address the issue of extracting multiple-item features, collaborative filtering recommendation has emerged as an alternative approach [21]. The core principle of collaborative filtering is similarity, which involves categorizing users and items based on their similarity and making recommendations accordingly. Collaborative filtering recommendation has found widespread application in various domains, including e-commerce [22], session recommendation [23], and article recommendation [24]. In collaborative filtering, two extensively studied methods are user-based collaborative filtering and item-based collaborative filtering [25]. User-based approaches predict user ratings based on the similarity of rating behaviors among users, while item-based approaches predict user ratings based on the similarity between predicted items and the actual items chosen by users. Due to the distinct principles of these two methods, their performances vary in different application scenarios. User-based recommendations tend to be more socialized, reflecting the popularity of items within the interest group to which the user belongs, while item-based recommendations are more personalized, reflecting the user's individual interest preferences.

In collaborative filtering recommendation, the cold-start problem is also present, and many researchers have proposed several solutions to address it. For example, Zhang et al. [26] identified the significant impact of different datasets, user attributes, the number of nearest neighbors, and the number of items on recommendation results. They developed an optimized user-based collaborative filtering recommendation system that tackles the issue of varying user rating scales by standardizing the original user data. By incorporating weighted user attributes and linear combinations with user rankings, they enhanced the overall user similarity. Experimental results demonstrated that this algorithm successfully mitigated the influence of cold-start problems and provided accurate recommendations.

2.3 Hybrid Recommendation

Due to the inherent limitations of individual recommendation algorithms, a hybrid recommendation has emerged as an approach to improve the overall recommendation performance by combining different models to complement their shortcomings [18, 19, 27]. Hybrid recommendation models blend two or more recommendation algorithms, thereby mitigating issues such as user cold-start and item cold-start, and overcoming the limitations of single algorithms.

For example, Zhang et al. [27] proposed a hybrid recommendation algorithm based on collaborative filtering and video genetics. The algorithm first constructs a user-item matrix, calculates user similarity, and performs clustering using k-means to generate a recommendation list. By analyzing the genetic structure of videos and combining style preferences and regional preferences, genetic preferences are formed. The weights of these genetic preferences are determined through linear regression. The objects are then ranked based on their degree of genetic preference, and the top-ranked objects are selected as the final recommendations. The algorithm combines the recommendation results from collaborative filtering and video genetics by assigning weights to each recommendation. However, they encounter issues such as high complexity and computational overhead.

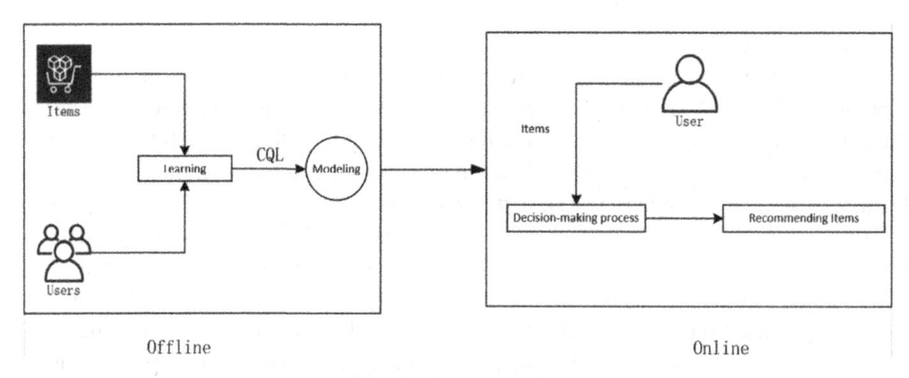

Fig. 1. System model

3 Preliminaries

Markov chain, named after the Russian mathematician Markov, is a mathematical model that represents a stochastic process. It is used to study the behavior and predict the future states of a system based on the current state, assuming that the future states only depend on the present state and are independent of the past states. This modeling technique is widely used in various fields, including probability theory, statistics, physics, computer science, and economics, to analyze and understand the dynamics of complex systems with probabilistic transitions.

Definition 1: Markov Chain. A Markov chain is a stochastic process that satisfies the property that the state of the system at time $t + 1$ depends only on the state at time t. It is characterized by the Markov property, which states that the future states are independent of the past states given the present state. In other words, the state transitions in a Markov chain are memoryless, and the probability of transitioning to a future state depends solely on the current state.

4 Construction of HMGR

Definition 2: Critical Queue Length (CQL). Once the user geolocation queue length reaches a specific threshold, the accuracy of the collaborative filtering recommendation algorithm begins to outperform that of the Markov chain recommendation.

The system model is depicted in Fig. 1. Initially, the system learns from a large dataset of user-profiles and check-in histories obtained from offline data. During the learning phase, two algorithms, namely the Markov chain algorithm and collaborative filtering recommendation algorithm, are applied to generate recommendations for users. By comparing the accuracy of these two algorithms, the system aims to determine the threshold point CQL where the Markov chain algorithm's accuracy starts to become lower than that of the collaborative filtering recommendation system. Subsequently, the system models this threshold point CQL and incorporates it into the decision-making process. When an online user interacts with the system and provides their check-in history, the system compares the length of the user's check-in history with the value of CQL. Based on this comparison, the system distinguishes between new users and existing users. If the length of the user's check-in history is less than or equal to CQL, the system identifies the user as a new user and employs the Markov chain algorithm to recommend items. Conversely, if the length of the user's check-in history exceeds CQL, the system recognizes the user as an existing user and utilizes the collaborative filtering recommendation algorithm to provide personalized recommendations.

4.1 Markov Chain in HMGR

Markov chains rely primarily on the current location of the user and the probabilities of transitioning between locations, rather than relying heavily on extensive historical data. In cold start scenarios, where there is limited data for personalized user modeling, Markov chains can utilize the available data to predict the user's next potential location and provide reasonably accurate recommendations. Therefore, Markov chains can be used to recommend geographical locations. Specifically, the predictive power of Markov chains can be leveraged to offer recommendations for new users.

Below, we present the pertinent definitions of Markov chains in our HMGR model.

Definition 3: State Pair. A state pair refers to the transition from one geographical location state to another. It represents the change in the geographical location, where the transition from state l_i to state l_j is denoted as $l_i \rightarrow l_j$.

Definition 4: State Transition Probability. In the context of geolocation, a transition from one geolocation state to another is referred to as a geolocation state transition. This process captures the relationship between state transitions and time as geolocation states change over time. The likelihood of a geolocation transitioning from one state to another over a specific time period is quantified as the state transition probability, as illustrated in Formula 1.

$$P_{ij} = \frac{C(l_i \to l_j)}{C} \tag{1}$$

where $C(l_i \to l_j)$ represents the number of occurrences in which the geographical location transitions from state i to state j in the user's interaction with the geographical location queue, while C represents the total number of transitions for state l_i.

Definition 5: Transition Probability Matrix. In the context of user geographical location states, where there are n possible states denoted by $\{l_1, l_2, l_3, \cdots, l_n\}$, the transition probability from state l_i to state l_j is represented as p_{ij}. These transition probabilities are combined to form a transition probability matrix, as shown in Formula 2.

$$P = \begin{pmatrix} p_{11} & \cdots & p_{1n} \\ \vdots & \ddots & \vdots \\ p_{m1} & \cdots & p_{mn} \end{pmatrix} \tag{2}$$

Typically, a standard Markov chain model can be represented using a triplet. However, in this paper, we employ a modified Markov chain model that is represented using a quadruplet, as depicted in Formula 3.

$$MC < L, P, M_t, Q > \tag{3}$$

where L represents the set of states in the model, denoted as $L = \{l_1, l_2, l_3, \cdots, l_n\}$, where n represents the number of states. P represents the state transition matrix, denoted as $P = [p_{ij}]_{n \times n}$. The element p_{ij} represents the probability of transitioning from state l_i at time t to state l_j at time t + 1. M_t represents the probability distribution of a user's state at time t, denoted as $M_t = \{m_1, m_2, m_3, \cdots m_n\}$, where m_i is the probability of being at l_i. Q represents the set of geographical locations that the user has not visited before, denoted as $Q = \{q_1, q_2, q_3, \cdots, q_m\}$, where m represents the set of unvisited locations.

The performance of a Markov chain recommendation system relies on the accurate calculation of state transition probabilities. If the state definitions are not precise or there are errors in the calculation of transition probabilities, it may lead to a decline in the system's performance. Therefore, the calculation of state transition probabilities is crucial. In our HMGR model, we use a typical Markov chain that utilizes the state transition matrix based on the current geographical location at time t to predict the user's geographical location at time $t + 1$. This approach helps us address the cold start problem, where limited data for personalized user modeling is available. By computing the probability distribution M_t for the user's current geographical location, we can use Formula 4 to calculate the state transition probability distribution M_{t+1} for the user at time t + 1.

$$M_{t+1} = M_t \times P \tag{4}$$

In our HMGR model, the Markov chain begins by collecting the user's historical geographical location data and organizing it into a time series. Each specific location within the sequence is treated as a state. By analyzing the user's historical sequence of geographical locations, we calculate the probabilities of transitioning between adjacent locations, capturing the likelihood of moving from one location to another. It's worth noting that our approach utilizes a one-step Markov chain, where each geographic location transition occurs in a single step from one location to another. This simplification allows us to effectively model and predict user behavior in the context of geographic location recommendations. Finally, we compute the probability distribution M_{t+1} and compare it to the set Q of geographical locations the user has not visited yet, thereby predicting the user's geographic location in the next time step.

The following Example 1 demonstrates the recommendation process using Markov chains in HMGR.

Table 1. User behavior trajectory.

User	Trajectory
u_1	$l_1 \rightarrow l_2 \rightarrow l_3 \rightarrow l_1 \rightarrow l_3$
u_2	$l_2 \rightarrow l_3$
u_3	$l_1 \rightarrow l_2 \rightarrow l_1$

Example 1: As shown in Table 1. In the table, u_1, u_2, u_3 represents three users, $L = \{l_1, l_2, l_3\}$ represents three sets of geographic locations, and $l_1 \rightarrow l_2$, $l_2 \rightarrow l_3$, $l_3 \rightarrow l_1$, $l_1 \rightarrow l_3$ represents the state pairs. According to Formula 1, we can calculate $p_{12} = 2/3$, $p_{13} = 1/3$, $p_{21} = 1/3$, $p_{23} = 2/3$, $p_{31} = 1$ and $p_{32} = 0$. Therefore, we obtain the state transition matrix $P = \begin{pmatrix} 0 & 2/3 & 1/3 \\ 1/3 & 0 & 2/3 \\ 1 & 0 & 0 \end{pmatrix}$.

Here, we set the probability of transitioning to the current location as zero. After obtaining the transition probability matrix, we can calculate the initial state probability distribution for the geographic location of user u_2 at time t. Given that the geographic location of user u_2 at time t is l_3, the initial state probability distribution of the user's state as $M_t = (0\ 0\ 1)$. Next, we can calculate the geographic location state transition probabilities for $M_{t+1} = (1\ 0\ 0)$ using Formula 4.

Upon analyzing the results, it becomes evident that the HMGR model employs a Markov chain to predict user u_2's level of interest in different geographical locations. The model indicates an interest level of 1 for location l_1, 0 for location l_2, and 0 for location l_3. Furthermore, when compared to the set $Q = \{l_1\}$, the prediction suggests that the most probable destination for user u_2 in the next stage is l_1.

4.2 Collaborative Filtering in HMGR

In cases where users have limited historical data, Markov chains can provide relatively accurate recommendations. Markov chains primarily rely on the current location of the user and the transition probabilities between locations, rather than relying heavily on extensive historical data. This enables Markov chains to predict the user's next potential location using the available data and offer reasonably accurate recommendations. However, the Markov chain model may not fully capture user preferences and personalized needs, as it only considers transition probabilities and lacks direct consideration of user interests and other contextual information. On the other hand, collaborative filtering algorithms can leverage user similarities and abundant historical behavior data to predict locations that users may like. Therefore, when users have a rich amount of historical data, collaborative filtering recommendations usually achieve higher recommendation accuracy.

In traditional collaborative filtering, there is a bias in the recommendation results due to imbalanced user interactions with geographical locations. The interactions tend to favor highly active users and popular locations, leading to a biased recommendation. The overwhelming interactions from active users and popular places overshadow other potentially relevant locations. This bias undermines the diversity of recommendations, making the recommendation system resemble more of a search engine rather than a personalized system.

A User-IIF algorithm [28] has been proposed in the literature to effectively address the issue of popularity bias, where popular items tend to become even more popular while less popular items are overlooked. In the context of geographical location recommendation, it is assumed that users exhibit similar behavior towards different locations. In our HMGR model, we have made slight modifications to its definition in order to mitigate the impact of popular locations. The algorithm introduces a modification to the user similarity calculation. We have further refined its definition to eliminate the influence of popular geographical locations. The calculation formula is presented as Formula 5.

$$w_{ij} = \frac{\sum_{l \in N(i) \cap N(j)} \frac{1}{\ln(1+N(l))}}{\sqrt{N(i) \times N(j)}} \tag{5}$$

where $N(i)$ represents the set of geographical locations that user i has provided positive feedback on. $N(j)$ represents the set of geographical locations that user j has provided positive feedback on. l represents the set of geographical locations for which both user i and user j have provided positive feedback. $N(l)$ represents the set of users who have given positive feedback on the geographical locations in set l.

After obtaining the similarity between users, the collaborative filtering recommendation algorithm recommends the geographical locations visited by the K most similar users to the target user. This recommendation is determined using Formula 6.

$$Score(u, i) = \sum_{v \in S(u,k) \cap N(i)} w_{ij} \times r_{vi} \tag{6}$$

where $S(u, k)$ represents the top k users who have the closest interests to user u. $N(i)$ is the set of users who have interacted with geographical location i. w_{uv} denotes the similarity of interests between user u and user v. r_{vi} represents the interest of user v in geographical location i. Since single-action implicit feedback data is used, all r_{vi} values are considered as 1.

The recommendation process of collaborative filtering is as follows: Firstly, collect the historical behavior data of users regarding geographical locations. Then, calculate the similarity between users to select the top K users who are most similar to the target user. These selected users are referred to as candidate users. Finally, based on the degree of association between the candidate users and the target user, considering factors such as locations already viewed by the target user, sort and filter the recommendation results to obtain the final recommendations.

The following Example 2 demonstrates the recommendation process using collaborative filtering in HMGR.

Example 2: We continue to use the data from Table 1 to predict the next geographical location that user u_3 is likely to visit. After calculating the similarity between users using Formula 5, we obtained the following similarity values: $w_{31} = 0.67$, $w_{32} = 0.29$. In this example, assuming $K = 1$, the candidate user is u_1. By comparing the historical data of users u_2 and u_3, we found that user u_2 has previously visited location l_3, while user u_3 has not visited l_3 yet. Next, using Formula 6, we calculated the $Score(u_3, l_3) = 0.67$, and therefore, we recommend geographical location l_3 to user u_3.

5 Experiments

In the experiment, we conducted a comparative study of collaborative filtering recommendation, Markov Chain recommendation, and our HMGR recommendation model using evaluation metrics such as accuracy, recall rate, and F-value. These solutions were implemented in Java language on a personal computer equipped with an AMD Ryzen 7 5800H CPU and 16GB RAM. We utilized the Foursquare dataset, which contains offline data for location recommendation. Specifically, we selected user geographical location interaction sequences from different areas of New York, USA. 80% of the data was used as the training set, and 20% of the data was used as the test set, for constructing the user similarity matrix and conducting recommendation evaluations.

Let R_u represent the list of recommended geographical locations calculated by the model based on user behavior in the training set, and T_u represent the list of geographical locations that the user will actually visit in the future based on the test set. The primary evaluation method is as follows:

(1) Precision measures the proportion of accurately recommended geographical locations among the samples predicted as other locations. This metric calculates the ratio of correctly classified positive samples to the total number of samples classified as positive by the classifier. It indicates how many locations in the predicted recommendation list are actually visited by the user in the future. The precision is calculated using Formula 7.

$$Precision = \frac{\sum_{u \in U} R_u \cap T_u}{\sum_{u \in U} R_u} \tag{7}$$

(2) Recall measures the proportion of correctly recommended locations among the recommended results. This metric reflects how many of the locations that the user will actually visit in the future are accurately predicted by the recommendation algorithm. It is calculated by dividing the number of correctly classified positive samples by the total number of actual positive samples. The recall can be computed using Formula 8.

$$Recall = \frac{\sum_{u \in U} R_u \cap T_u}{\sum_{u \in U} T_u} \tag{8}$$

(3) The F1 score, also known as the balanced F-score, is a metric that balances both precision and recall, providing an overall measure of the model's performance. It can be seen as the harmonic mean of precision and recall. The F1 score is calculated using Formula 9.

$$F1 = \frac{2 \times Precision \times Recall}{Precision + Recall} \tag{9}$$

5.1 Determining the Recommendation Length L

Based on Fig. 2, Fig. 3 and Fig. 4, the HMGR model demonstrates improved accuracy compared to the individual collaborative filtering and Markov chain models. The recall rate of the HMGR model initially falls between the collaborative filtering and Markov chain recommendations, but as the number of recommended locations increases, the HMGR model surpasses both collaborative filtering and Markov chain recommendations. The F1 score of the HMGR model consistently falls between collaborative filtering and Markov chain recommendations. As the number of recommended locations increases, the accuracy also improves. Therefore, we set the recommendation length to 5. When the recommendation length is 5, the F1 score of our HMGR model is slightly lower than that of the collaborative filtering, but this difference can be considered negligible.

As the recommended length increases, the decline in precision, recall, and F1 score can be attributed to the presence of varying quality of recommended locations within longer recommendation lists. Lower-quality recommendations may be mixed with high-quality ones, which can adversely affect the precision and recall of the recommendations. The F1 score serves as a comprehensive metric that takes into account both precision and recall, providing an assessment of the overall model performance. When precision and recall decrease, the F1 score naturally decreases as well.

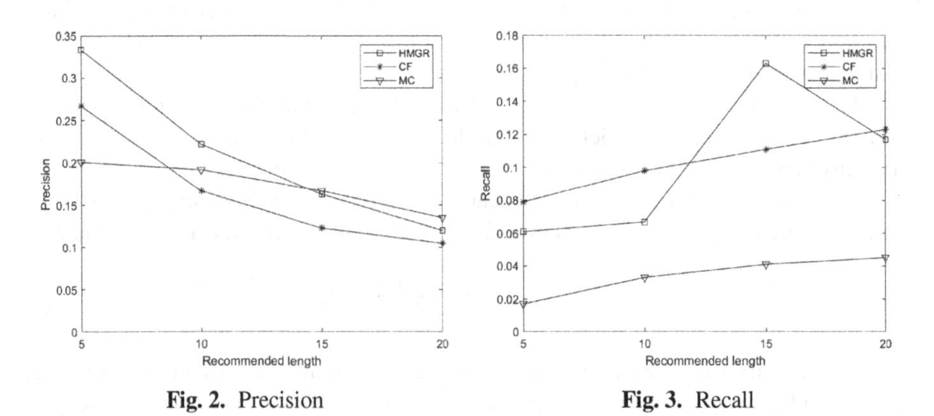

Fig. 2. Precision **Fig. 3.** Recall

5.2 Solution to the Cold-Start Problem

As shown in Fig. 2, our proposed HMGR model achieves significantly higher precision compared to the Markov Chain and Collaborative Filtering recommendations when the recommendation length is set to 5. This indicates that our HMGR model effectively addresses the cold-start problem, where limited user data is available for accurate recommendations. By combining the strengths of the Markov Chain and Collaborative Filtering approaches, the HMGR model provides more accurate and personalized recommendations even in scenarios with sparse user history.

In the HMGR model, we combine the Markov chain and collaborative filtering recommendation methods. When the user's historical data is limited, the Markov chain can provide relatively higher recommendation accuracy. The Markov chain primarily relies on the transition probabilities between the user's current and future locations, rather than relying on extensive historical data. In such cases, the Markov chain can effectively predict the user's next likely location and offer accurate recommendations. However, the Markov chain model has limitations in capturing user preferences and personalized needs since it only considers transition probabilities and overlooks user interests and contextual information. On the other hand, collaborative filtering algorithms can leverage the similarity between users and their extensive historical behavior data to predict locations that a user is likely to prefer. Thus, when users have abundant historical data, collaborative filtering algorithms generally provide higher recommendation accuracy. Our HMGR model combines both approaches, leading to improved accuracy in recommendations. By leveraging the strengths of both the Markov chain and collaborative filtering, we can provide more accurate and personalized recommendations to users, overcoming the limitations of each individual method.

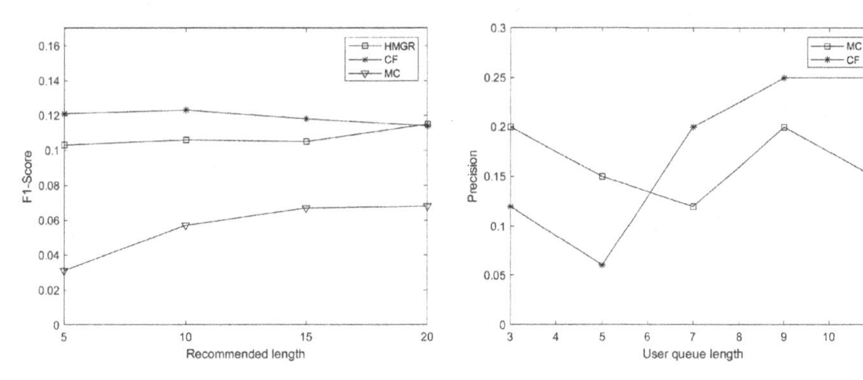

Fig. 4. F1-score

Fig. 5. User queue length

5.3 The Determination of New Users

As shown in Fig. 5. When the length of the user queue is less than 6, the accuracy of the Markov chain is higher than that of collaborative filtering. However, when the length exceeds 6, the accuracy of the Markov chain is lower than that of collaborative filtering. This suggests that for users with a relatively short history of interactions (queue length < 6), the Markov chain approach performs better in terms of recommendation accuracy. On the other hand, for users with a longer history of interactions (queue length > 6), collaborative filtering yields higher accuracy in recommendations. Therefore, the HMGR model takes into account the length of the user queue to determine whether a user is considered new or existing and accordingly selects the appropriate recommendation method to provide accurate and personalized recommendations.

This is because the Markov chain recommends geographic locations to users based on probabilistic transitions, without directly considering user interests. As the user queue length increases, the accuracy of the Markov chain does not necessarily improve. On the other hand, collaborative filtering recommendation utilizes users' historical visit data to observe their preferences and the likelihood of revisiting specific geographic locations. As the user queue gradually grows, the user similarity matrix becomes more refined, resulting in increased accuracy. When determining whether a user is new, we can compare the user queue length at which the accuracy of collaborative filtering recommendation surpasses that of the Markov chain through testing.

6 Conclusion

In this paper, we present a novel and efficient hybrid recommendation algorithm, the HMGR model. By combining the Markov chain algorithm with collaborative filtering, the HMGR model successfully overcomes the limitations of traditional recommendation methods, especially in terms of handling cold-start issues and providing personalized recommendations. The experimental results demonstrate that the HMGR model achieves significant improvements in various scenarios and exhibits outstanding performance in terms of accuracy and efficiency. Of particular significance is the HMGR model's innovative approach to address the cold-start problem in collaborative filtering recommendations. By leveraging the Markov chain algorithm to predict users' latent interests, the HMGR model can deliver effective recommendation services even for new users in the system. For emerging recommendation systems, solving the cold-start problem is critical, and the remarkable performance of the HMGR model brings renewed hope in this aspect. However, we also acknowledge that there is still room for further refinement in the HMGR model. Future research could explore more intricate Markov chain models to enhance the accuracy of predicting changes in users' interests. Additionally, integrating other recommendation algorithms into the HMGR model presents an enticing avenue for further enhancing the recommendation system's overall performance.

In conclusion, this study firmly establishes the effectiveness and potential of the HMGR model in the user geolocation recommendation system. By blending the power of the Markov chain and collaborative filtering methods, we successfully address the cold-start problem and substantially improve the recommendation system's accuracy. We believe that this research will offer valuable insights for the ongoing development and enhancement of recommendation systems, providing users with more personalized and precise geolocation recommendation services.

Acknowledgement. This research is supported by the National Natural Science Foundation of China (No. 61962029, No. 62062045, No. 62262033), the Jiangxi Provincial Natural Science Foundation of China (No. 20202BAB212006) and the Science and Technology Research Project of Jiangxi Education Department (No. GJJ201832).

References

1. Tran, D.T., Huh, J.H.: Building a model to exploit association rules and analyze purchasing behavior based on rough set theory. J. Supercomput. **78**(8), 11051–11091 (2022)
2. Gupta, M., Thakkar, A., Gupta, V., Rathore, D.P.S.: Movie recommender system using collaborative filtering. In 2020 international conference on electronics and sustainable communication systems (ICESC), pp. 415–420. IEEE (2020)
3. Zhang, C., Mei, Z., Wu, B., Zhao, Z., Yu, J., Wang, Q.: Query with assumptions for probabilistic relational databases. Tehnički vjesnik **27**(3), 923–932 (2020)
4. Zhang, C., Mei, M., Mei, Z., Zhang, J., Deng, A., Lu, C.: PLDANet: reasonable combination of PCA and LDA convolutional networks. Int. J. Comput. Commun. Control **17**(2), 1–13 (2022)
5. Wu, B., Chen, X., Wu, Z., Zhao, Z., Mei, Z., Zhang, C.: Privacy-guarding optimal route finding with support for semantic search on encrypted graph in cloud computing scenario. Wirel. Commun. Mob. Comput. **2021**, 1–12 (2021)
6. Guo, X., Yin, S.C., Zhang, Y.W., Li, W., He, Q.: Cold start recommendation based on attribute-fused singular value decomposition. IEEE Access **7**, 11349–11359 (2019)
7. Papadakis, H., Papagrigoriou, A., Kosmas, E., Panagiotakis, C., Markaki, S., Fragopoulou, P.: Content-based recommender systems taxonomy. Found. Comput. Dec. Sci. **48**(2), 211–241 (2023)
8. Afoudi, Y., et al.: Impact of feature selection on content-based recommendation system. In: Proceedings of the 2019 International Conference on Wireless Technologies, Embedded and Intelligent Systems (WITS), pp. 1–6 (2019)
9. Yadalam, T.V., Gowda, V.M., Kumar, V.S., Girish, D., Namratha, M.: Career recommendation systems using content based filtering. In 2020 5th International Conference on Communication and Electronics Systems (ICCES), pp. 660–665. IEEE (2020)
10. Javed, U., Shaukat, K., Hameed, I.A., Iqbal, F., Alam, T.M., Luo, S.: A review of content-based and context-based recommendation systems. Int. J. Emerging Technol. Learn. (iJET) **16**(3), 274–306 (2021)
11. Fu, M., Qu, H., Yi, Z., Lu, L., Liu, Y.: A novel deep learning-based collaborative filtering model for recommendation system. IEEE Trans. Cybernet. **49**(3), 1084–1096 (2018)
12. Afsar, M.M., Crump, T., Far, B.: Reinforcement learning based recommender systems: a survey. ACM Comput. Surv. **55**(7), 1–38 (2022)
13. Shokeen, J., Rana, C., Rani, P.: A trust-based approach to extract social relationships for recommendation. In: Khanna, A., Gupta, D., Pólkowski, Z., Bhattacharyya, S., Castillo, O. (eds.) Data Analytics and Management. LNDECT, vol. 54, pp. 51–58. Springer, Singapore (2021). https://doi.org/10.1007/978-981-15-8335-3_6
14. Shokeen, J., Rana, C.: A trust and semantic based approach for social recommendation. J. Ambient. Intell. Humaniz. Comput. **12**(11), 10289–10303 (2021)
15. Shokeen, J., Rana, C.: Social recommender systems: techniques, domains, metrics, datasets and future scope. J. Intell. Inform. Syst. **54**(3), 633–667 (2020)
16. Walek, B., Fajmon, P.: A hybrid recommender system for an online store using a fuzzy expert system. Expert Syst. Appl. **212**, 118565 (2023)
17. Chen, Z., Zhu, S., Niu, Q., Zuo, T.: Knowledge discovery and recommendation with linear mixed model. IEEE Access **8**, 38304–38317 (2020)
18. Walek, B., Fojtik, V.: A hybrid recommender system for recommending relevant movies using an expert system. Expert Syst. Appl. **158**, 113452 (2020)
19. Berkani, L., Hanifi, R., Dahmani, H.: Hybrid recommendation of articles in scientific social networks using optimization and multiview clustering. In: Hamlich, M.., Bellatreche, L., Mondal, A., Ordonez, C. (eds.) SADASC 2020. CCIS, vol. 1207, pp. 117–132. Springer, Cham (2020). https://doi.org/10.1007/978-3-030-45183-7_9

20. Khanal, S.S., Prasad, P.W.C., Alsadoon, A., Maag, A.: A systematic review: machine learning based recommendation systems for e-learning. Educ. Inf. Technol. **25**, 2635–2664 (2020)
21. Meng, D.: Collaborative filtering algorithm based on trusted similarity. In: Proceedings of the 2018 IEEE 3rd International Conference on Signal and Image Processing (ICSIP), pp. 572–576 (2018)
22. Jebamalar, B., Sudha, K., Maheswari, B., Nalini, M., Dinesh, M.G., Subramanian, R.S.: Collaborative filtering for bargaining and customer feedback in e-commerce. In: 2023 7th International Conference on Intelligent Computing and Control Systems (ICICCS), pp. 1804–1808. IEEE (2023)
23. Chen, Z., et al.: Towards explainable conversational recommendation. In: Proceedings of the Twenty-Ninth International Conference on International Joint Conferences on Artificial Intelligence, pp. 2994–3000 (2021)
24. Nair, A.M., Benny, O., George, J.: Content based scientific article recommendation system using deep learning technique. In: Suma, V., Chen, J.I.-Z., Baig, Z., Wang, H. (eds.) Inventive Systems and Control. LNNS, vol. 204, pp. 965–977. Springer, Singapore (2021). https://doi.org/10.1007/978-981-16-1395-1_70
25. Muhammad, M., Rosadi, D.: Comparison of user-based and item-based collaborative filtering methods in recommender system. In: AIP Conference Proceedings, vol. 2720, no. 1. AIP Publishing (2023)
26. Zhang, J., Lin, Z., Xiao, B., Zhang, C.: An optimized item-based collaborative filtering recommendation algorithm. In: 2009 IEEE International Conference on Network Infrastructure and Digital Content, pp. 414–418. IEEE (2009)
27. Zhang, W., Wang, Y., Chen, H., Wei, X.: An efficient personalized video recommendation algorithm based on mixed mode. In: 2019 IEEE International Conferences on Ubiquitous Computing & Communications (IUCC) and Data Science and Computational Intelligence (DSCI) and Smart Computing, Networking and Services (SmartCNS), pp. 367–373. IEEE (2019)
28. Breese, J.S., Heckerman, D., Kadie, C.: Empirical analysis of predictive algorithms for collaborative filtering. arXiv preprint arXiv:1301.7363 (2013)

Non-essential Perspective on Thinking Law of Data Protection and Utilization

Zhao Li, Honggui He(✉), Wen Wang, Siqi, and Tao

Law School, Jiangxi University of Finance and Economics, Nanchang 330013, Jiangxi, China
Johnleehust@hotmail.com

Abstract. The protection and utilization legal system of data is of fundamental importance in contemporary digital age. Existing research a have a problem of essentialism of theoretical thought, which hinders researchers and legislators from exploring the institutional framework that matches the reality of digital society according to the inherent differences of data concepts, the process of data practice, the application and value tradeoffs in this process. On the basis of criticizing the essentialism perspective taken by recent studies, this article takes non-essential Perspective for thinking legal system of data protection and utilization, and pursuits creative data utilization. It proposes a data usufruct operating within an expanded framework of right to human dignity, which might provide a new way of thinking data protection-utilization law that could balance the basic morality of the digital age and the national digital economic policy.

Keywords: Data protection–utilization · Essentialism · The management of data value production · Extended personality rights

1 Introduction

The "protection-utilization" legal system of data is the pillar of the legal governance system of digital society. People often divide the protection of data rights and interests from the rational use of data, but the two are actually the whole. It is impossible to separate the rules that protect the rights and interests of data from the rules of data sharing and utilization. The "protection-utilization" legal system of data is a rule system that discards the fundamental difference between the protection of data rights and interests and the rational use of data and provides rule of law support for data flow in the context of digital economy. The study of "protection-utilization" legal system of data is the overall investigation of this rule system.

At present, the research on the "protection-utilization" legal system of data is limited by the institutional logic of essentialism, so it is difficult to respond to the complexity of data practice. As a mode of knowledge production, "essentialism" refers to the cognitive path of obtaining conclusions, judgments or other types of knowledge by applying this "essence" to concrete experience from the "essence" of universal sharing of things. Existing studies often put the data under the existing institutional framework by defining the essential characteristics of the data. The institutional logic of "protection-utilization"

S. Zhang et al. (Eds.): BigData 2023, LNCS 14203, pp. 63–79, 2023.
https://doi.org/10.1007/978-3-031-44725-9_5

of data shows a strong tendency of essentialism. However, it is difficult for researchers to deduce "how to protect and utilize data" on the basis of scientific understanding of "what data is".

In order to avoid the solidification of thinking caused by the institutional logic of essentialism, we need to realize that the essence of data is not in itself, but in the process of utilization. In the face of the interest pattern created by the use of data, it is considered that the legal system should proceed from the legal interests and distribute rights, obligations and responsibilities among different subjects. The legal interests of "protection-utilization" legal system of data are constructed from the evaluation of data value production. To define the legal interests arising from the use of data is to manage the production of data value. On the basis of managing data value production, construct a "data usefulness legal system that operates within the expanded personality protection framework and encourages innovative value production", or can respond to the legal problem of constructing "protection-utilization" legal system of data.

2 Research Objects and Basic Concepts

The protection and utilization of the data contains the two aspects of value, one is the individual value that data protection points to, for instance, the protection of personal privacy, whether it regards privacy as a free right not to be peeped by the government or society, or as the individual's freedom to choose and act independently, in essence, it points to the protection of individual values. The second is the economic value pointed by data. Economic life brought by the development and prosperity of the rich material life greatly promote the political, cultural, legal, and so on various aspects of social life of prosperity, but when too pursuit of economic value, or absolute economic value, what it brings is that capital operates out of control in accordance with its logic, which is bound to erode other social values, that is, the values of other social life will be forced to make concessions and economic values. It includes the value of the individual. Individual value and economic pursuit are in an either-or situation, but this is not what a good social development expects. Therefore, it is necessary to build a new value environment-the pursuit of multiple values, limit the disorder of utilization through protection, and promote better protection through utilization. In the multiple value evaluation system, whether to achieve the protection of individual value or the pursuit of economic value, the goal is the same, in order to pursue creative value and integrate the two or even multiple values to achieve the construction of multiple values.

Currently, a lot of studies on the "protection - utilization" legal system of data try to project its functions to the existing institutional framework by defining the essential characteristics of the data. This path of "essentialism" does not apply to the complex problems faced by the "protection-utilization" legal system of data. Before criticizing the essentialist institutional logic of "protection-utilization" of data, we need to recognize the characteristics and limitations of essentialism and its institutional logic itself. Essentialism has at least three basic presupposition: (1) the same type of things exist by some common attributes (which he attributes is incidental), is an essential part of naming, alleged or define things [1]; (2) the attribute of things there is a "hard core" of the content, will not be different with historical process and the situation changes; (3)

as to the nature of property or other let a priori determines the specific presentations in the experience things to rule out the unique needs of thinking "application".

Although essentialism is an important way for people to understand the world, society and self, its effectiveness as a mode of knowledge production has been criticized by many scholars. First of all, some scholars do not believe that the concept has a general meaning. Second, the essence of things is not constant, and the "hard core" content that advocates excluding change is actually out of the wrong understanding of the concept of essence. Finally, the essential stipulation of empirical judgment is also challenged. Thus, it can be seen that the three basic presuppositions of the essentialist mode of knowledge production all have inherent limitations at the philosophical level.

3 Problems and Current Situation

As an important way of knowledge production in the study of law, essentialism can effectively maintain a stable and logical legal system, but it will also lead to the mechanical rigidity of legal practice and cannot meet the practical needs of data governance [2]. The "protection-utilization" legal system of data based on the concept, nature, basic principles and specific framework of data is inherent because it involves three propositions related to essentialism-- the universal implication of the concept, the self-referential nature of legal elements, and the inherent stipulation of the legal system.

The institutional logic based on the concept of data and its limitations. The essentialist institutional logic of data "protection-utilization" is reflected in the following belief of some scholars, that is, if the connotation of data is not defined in a scientific way, it cannot be effectively regulated at the legal level. The understanding of data in legal research mainly includes "bit theory", "file theory" and carrier theory, only taking the carrier theory as an example. This theory defines data as a bit form represented by a combination of 0 and 1 on the basis of binary circulating on computers and networks [3]. Jurists choose this definition mainly to distinguish it from the information stored in traditional media, in order to take the data stored and processed by computer as a unique object of study. Scholars seem to explore its legal regulation from the nature and definition of data, but in fact they realize that there is a profound difference between the "protection-utilization" of data in the computer context and the privacy protection in the non-computer background. Thus, the former is treated as a new problem under the current technological conditions. Therefore, the clarification of the connotation of data science is not the logical starting point of "protection-utilization" legal system of data but is "selected" from many definitions according to the needs of legal research and legal evaluation. Thus, it can be seen that the essentialist institutional logic based on the scientific definition of data lacks a certain degree of authenticity. It is precisely because it conceals the real process of "selecting" the scientific connotation of data, the universal concept of data does not fully specify the different data that need to be treated differently in legislation. For the need of comprehensive regulation, a lot of legislation covers a wider range of objects in the definition of data, which conflicts with the theoretical starting point.

Institutional logic based on the nature of data and its limitations. The institutional logic has at least three limitations: first, the existing academic summary of the nature of the data cannot cover any possible situation, that is to say, not all data are "non-competitive" and "non-exclusive". Secondly, if we start from the "non-specificity" and "non-independence" of the data and deny the data as the object of civil law, this institutional logic will cover up the real process of data replication and evaluation. Finally, this essentialist institutional logic does not realize that it is the "application" that determines the nature of the data, not the nature of the data.

Institutional logic based on basic principles and its limitations. In addition to the scientific connotation and essential attribute of data, the institutional logic of essentialism may also stipulate the attribution of data rights on the premise of basic principles such as "labor empowerment theory" or "Coase theorem". Unlike Locke's property, people can use data without occupying the original data set by copying. Therefore, the demand of individualization or even exclusive possession is not the hard demand of using data, but more important to explore the institutional framework to promote the common prosperity of mankind through the use of data. Secondly, the application of labor empowerment theory out of context will cause researchers to ignore the real process of value production and trade-off. Without the analysis of the value production process, it is impossible to determine the applicability of the labor empowerment theory. Compared with land and other traditional property, the data does not have a relatively definite value production path. While it creates huge economic value, it also causes value derogation due to external negative effects such as consumption manipulation, privacy invasion, information cocoon house, human participation and the marginalization of decision-making [4]. Therefore, apart from the real process of value production and trade-off, the application of labor empowerment theory in an essentialist way may cover up the complexity of the problem. Finally, the institutional exploration with the labor empowerment theory as the logical starting point masks the key role of "application" in the process of theoretical construction.

The institutional logic starting from the specific legal framework and its limitations. In the face of data as a new item, the most common way of regulation is to bring it into the existing legal framework. As long as it is consistent with the constitutive elements of the regulatory object of the legal framework, the data can be regarded as the same kind of object into the scope of effectiveness of the specific legal framework. However, because legislators did not foresee the mode of production of data value in the digital age when formulating these legal frameworks, the coincidence of these constituent elements with data is only an accidental phenomenon. it is difficult to overcome the inherent limitations of essentialist institutional logic by adhering to the mode of thinking that meets the elements. On the one hand, large amounts of data can't really meet the constituent elements required by these legal frameworks, thus can not be incorporated into the protection of specific legal frameworks. On the other hand, the constituent elements of different legal frameworks may create unnecessary internal cuts to data sets. Secondly, the practice of anchoring to a single legal framework according to the guidelines of the constituent elements may obscure the complexity of the production process of data value. Finally, from the data protection proposition of trade secrets, we can see that the key to comparing data to trade secrets is not its constituent elements, but the application focus of weighing the production of data value. Along the path of essentialism, people tend to get caught up in the laborious work of screening a large number of unstructured data sets in experience with constitutive elements, and it is difficult to pay attention to more fundamental problems.

4 From Essentialism to Non-essential Perspective of Protection and Utilization of Data

As an important way of knowledge production in legal research, essentialism can effectively maintain a fairly stable and logical legal system, but it will also lead to the mechanical rigidity of legal practice and cannot meet the practical needs of data governance. [2].

The criticism of essentialism does not mean moving towards the position of anti-essentialism. This will destroy people's perceptions about the "protection-utilization" of data and push the legal system into an abyss of total uncertainty [5]. The serious criticism of the logic of the essentialist system is based on the needs of practical seeking to break the old, fixed formula in the formation of legal knowledge, so that people can focus on the practical form pointed to the "protection-utilization" legal system of data. The digital economy rooted in data "protection-utilization" is a creative economy, and the development of data value does not follow the past stable model. This creative value production has led to the continuous changes in the interest pattern of all parties and possible legal status. Starting from existing legal frameworks, it is either difficult to support the creative nature of the digital economy, or it is impossible to respond to complex and diverse situations. In view of this, the research "protection-utilization" legal system of data needs to anchor a new foundation to meet the needs of researchers to judge legal issues in the context of a creative economy. The institutional logic of critical essentialism is to establish this foundation or practice form: the production management of data value.

4.1 Utilization of Data: A New Perspective

Starting from the use of data is the first step to break away from the essentialist way of thinking and reconstruct the institutional logic of "protection-utilization". According to the above analysis, to think about the "protection-utilization" legal system of data from the nature of data is to make the fresh data in the process of utilization into a "mummy of data concept" [6]. This seems to be the worship and worship of the nature of data science by legal people, but it actually blots out the life shown in the process of data utilization. For example, many legal researchers believe that in order to study the ownership of data, it is necessary to clarify the difference between data and information: the exercise of similar portability belongs to the data problem, while the platform collects personal information without the consent of the user. There is a certain theoretical basis for this distinction, and "data issues" pay attention to the security of the system and the integrity of data; and "information issues" focus on reducing the scope of being known and circulating for personal information. However, this "basis" cannot be understood according to the nature of the data itself, but can only be regarded as supported, changes, or hindered some kind of data utilization, and a certain subject is necessary. In other words, the real difference is not the essence of data, but the practical form of data utilization. For the "protection-utilization" legal system of research data, paying attention to data use is far more important than questioning its essence.

4.2 Production of Data Value: The Necessary Background

Starting from the real-life process of human material production that ensures that people can produce from the social structure [7], reconstruct the institutional logic of "protection-utilization" of data. This investigation refused to preset the premise of essentialism, but from the reality before. This premise is people. In this perspective, data utilization is not an abstract activity, but an activity that is understood in the relationship between data and people themselves. People are constantly customizing the use of data according to their own standards, and their experiences with data and the real world are also shaping the people as subjects and their scales. Thus, Data utilization is a process of constantly converging the use of data with human needs, that is, the process of producing its value. The people concerned by the "protection-utilization" legal system of data are the realistic people who live in the society. This requires researchers' thinking must be based on the social background of realizing value production through transactions and systematic operations, and keep two insights in mind: (1) Some value mode of production in the socialization process deviation due to distorted human scale direction;(2) Differential different value production methods compete for dominant position in the social field. The "protection-utilization" legal system of data inevitably affects the distribution pattern of data value production methods.

In the most universal sense, the production can be divided into two conflicting ways: people-oriented multi-value production and economic efficiency-centered value production. The former uses "data utilization" as a means to enhance human cognition, judgment and action, and promote human development and prosperity. Because of the diversity of human existence, the use of people-oriented data requires the pursuit of different social, aesthetic or ethical values according to individual needs in different situations. The "fuzziness" and "friction" which inevitably lead to inefficiency in the use of data are essential to this value mode of production in some cases [8]. Under the social conditions where economic efficiency-centered value production occupies a dominant position in most fields, people-oriented value production is in jeopardy. However, the technical design of encryption technology and embedded value still tries to enable people to better control their own information and data by changing the mode of economic operation and technical conditions. And set up the final line of defense for people to control their lives in the control network of capital, technology and power [9].

On the other hand, although the value production centered on economic efficiency takes digital products and services as the basis to meet human needs, it does not regard this satisfaction as the goal of data utilization. The main manifestation of this value mode of production is that institutions with equipment, large databases and professional data analysis technology mine commercially valuable information in order to improve products and services. Although the economic value production of data is the social labor of users, the analysis of data engineers, and the operation and maintenance of platform corporate employees, the products and services generated in the final analysis are not available for workers and wider public, so beginning to the existence of dissidents [10]. With the development of data processing and the prosperity of platform economy, digital capital gradually takes the lead in the whole process of data value, even by constructing false digital demand, to control users as producers of data value and consumers of data goods from the material level to the spiritual level. This kind of economic efficiency-centered data value production and people-oriented data utilization compete with each other, which together constitute the necessary background for understanding the "protection-utilization" legal system of data.

4.3 Managing Data Value Producing: The Application Focus

The "protection-utilization" legal system of data cannot only stay at the level of understanding the social process of data value production. Instead, it must be settled to a constructive and evaluated application level. As an integral part of cultural practice, the legal system cannot be satisfied with the scientific understanding of the realistic order but should integrate "interest" and "knowledge" [11]. challenge the realistic order by standardizing the existing interest pattern. The historical legitimacy of the values guaranteed and pursued by law such as justice, freedom and goodness lies in the contents that have not yet been realized in the realistic order [12]. Therefore, the "protection-utilization" legal system of data can only make a diagnosis and choice between consolidating the order dominated by digital capital and power and promoting the ideal prospect of a digital society that promotes the development of multiple endowments of human beings. And take the regulation of each data utilization behavior as an intervention in the real order, in order to construct a basic social order that meets the normative requirements in the context of the rapid development of digital technology. If we ignore the normative requirements of realizing individual autonomy, self-determination and subjectivity under the new conditions of digitalization and intelligence, and do not mention creating space for individuals to live in a fair, rational and transparent digital environment and realize the free development of personality, the personal data protection legal system is out of the question.

The application of "protection-utilization" legal system of data is to manage the production of data value, so that data utilization can produce desirable value combinations. The metaphor of management behavior means that the institutional "protection-utilization" legal system of data is regarded as an actor who promotes a specific value combination through judgment and decision-making in the ever-changing information current of complex world facts [13]. Data utilization supports a series of social processes full of liquid operation, such as flow, sharing, tracking, interaction, screen reading, remixing and so on. Legal evaluation needs to be the same dynamic and should be close to the management behavior that can respond to changes in situation. With the characteristics of liquid governance. The fact that data value production is creative means that the interest pattern may change beyond what legislators have imagined with data utilization, which requires "protection-utilization" not only to determine specific situations based on a priori rules, but to develop such rules from the reality created by data utilization actions. Norms can only be presented in the process of things [14], and the role of managers can highlight the importance of this fact more than the role of traditional legislators. Thus it can be seen that the starting point of thinking about the "protection-utilization" legal system of data is to regard society as a whole of data value production, and to support social practices that can produce desirable value combinations by means of empowerment, licensing, supervision and judicial adjudication in the dynamic process of data utilization [15].

As the focus of application, the production management of data value integrates the legislative activities, rule interpretation and other types of institutional practice of the "protection-utilization" legal system of data into a coherent system. The protection and utilization of data in the intelligent society is not a piecemeal or trivial problem, but involves the fundamental problem of maintaining human freedom, autonomy and autonomy through subjective construction under the new technological conditions. The "protection-utilization" legal system of data is not a governance tool for competing for special goals in independent battlefields, but a grand narrative supported by several scenarios. If "protection-utilization" cannot be regarded as a "meaningful whole", it will be impossible to construct a consistent legal system to respond to the omni-directional challenges posed by the digital society to human life at an invisible level. The application of data value production management can provide the ultimate significance for all the practices of the "protection-utilization" legal system of data. The integration of personality interests and property rights and interests, empowerment and supervision path, public law protection and private law protection, constitution and general legal framework is the key to bring digital utilization into the track of normative review and rule of law.

5 Human-Centered Innovative Value Production: The Ends

The management of data value production should be based on a belief in the legal practice of "protection-utilization" of the whole data, and [16] to prove the purpose, goal or principle of data value production management. It is hard to define this belief system from inside the legal system, and exploring this issue inevitably involves the debate on the basic principles of justice and the pursuit of values in the digital age. Before entering the discussion of the specific legal framework, it is necessary to go deep into the debate of political philosophy and clarify the belief basis of management data value production from four levels.

5.1 Maintaining Subjectivity of Human Beings

As the dominant technology category in the digital society, data utilization must first support the purpose of the technology itself. On the basis of eliminating all kinds of misunderstandings of modern social concepts, German philosopher Oswald Spengler traced technology back to the source of "survival strategy to support the will to power" [17]. As a kind of purposeful action, technology has its real existence only when it is produced creatively to counter the hostile environment. But as this action evolved from struggle and conquest with nature to plunder, enslavement and manipulation of the world, technology began to betray its purpose. Data may also be involved in the technological rebellion in new ways. As a typical way of contemporary data utilization, algorithmic governance does not regard people as practical, empirical, present individuals with complex emotions, but as fragments in a large number of scattered personal data in an atomized way. The architecture that provides the basic conditions for the use of such data focuses on the impersonal and fragmented collection of digital traces in daily life and communication only for the sake of prediction [18]. Individuals as moral actors or subjects are in jeopardy in this digital architecture that has evolved to support the widespread use of data.

Subjectivity is a normative category that limits the destructive effects of capital control and technological betrayal in modern society. In the aspect of law, subjectivity is expressed as the basic principles of "safeguarding human dignity" and "ensuring the free development of personality". Under the technological conditions of the digital society, the core of this personality interest is to "effectively defend against the possible personality intervention of others, so as to freely participate in social life and realize the free development of personality" The EU countries represented by Germany put the "right to self-determination of information" into the list of general personality rights, which is a typical way to protect subjectivity at the legal level [19]. Although, Although the human dominant position in the digital age can be maintained to a certain extent in the way of general personality rights and interests, the metaphor of "management" has the following additional value. (1) It can highlight the importance of value tradeoff. Even in the "inviolable areas" of "safeguarding private life", it is a tradeoff to regard the values of "inviolable freedom" and "human dignity" as absolute priority over other competitive principles [20]. The metaphor of management is more helpful to avoid treating the "protection-utilization" of data as a simple process in which the rules apply. (2) Can understand and meet the institutional needs of maintaining subjectivity from a broader level. The digital architecture can avoid the general personality rights protection model with the rule of informed consent as the core. Regarding "maintaining the status of human subjectivity without being weakened" as the management goal is conducive to flexibly mobilizing a variety of institutional resources, making legal governance embedded in the review and supervision of architectural design, and improving the situation of human freedom and dignity in the digital age as a whole.

5.2 Encourage Pluralistic and Innovative Utilization

On the premise that the status of human subjectivity will not be weakened and the realization of internal goals in the field of maintenance, it is necessary to consider whether data practice is conducive to innovating lifestyles in a diversified way. On the one hand, the use of diversity means that, first of all, it does not follow the way customized by the existing economic system (although it needs to rely on the economic system to cash in its mechanisms). Instead, the value of data is mined in the way prescribed in the fields of social, aesthetic, justice, culture and ethics. On the other hand, the creative use of data is not primarily about technology but refers to the innovative utility of data use: it can either release a way of life suppressed by the economic system or develop unprecedented life practices. These two aspects are inseparable, and the multiple use of data value can usually enrich people's social lifestyle. Promoting the common prosperity of mankind through the use of data is the ideal goal of the digital society [22].

If the use of data can be accompanied by the innovation of lifestyle, it should take precedence over the practice stipulated by the existing social and economic system in the tradeoff of the system of "protection-utilization" of data. This claim needs to be morally defended. Cybernetics and complexity theory are the weakest and most defensible paths. In the view of cybernetics theorists, the world as a whole obeys the second law of thermodynamics: chaos is increasing and order is decreasing. In areas where human society can make progress, the increase of local order can always be achieved [23]. In order to maintain the order of human life in a diverse environment, social systems must have the necessary diversity [24]. The innovation of life style is the source of complexity within human society as a system, in order to provide the necessary diversity to cope with the environment. However, the dominant structure of the existing social system, such as the business model, will always inhibit the innovation of people's way of life. The Internet and digital technologies make large enterprises more capable of defining and shaping people's needs and interests, and continue to promote the colonization of business models to the living world. Even if there is data utilization beyond the established model, as long as the dominant model or framework refuses to interact with it, data utilization with the potential for innovative lifestyles can be easily excluded from the social system [25]. As an important part of the social system regulator, the "protection-utilization" legal system of data should give priority to the use of data in innovative life styles in order to weaken the exclusion of social systems and promote the diversification of life styles.

5.3 Pursuing Economic Value: Benefits Most General Ones

On the premise of meeting the above three management objectives, the economic value created by the data utilization should benefit the most general public, especially those who are most disadvantaged in the digital society. Compared with the traditional market economic system which relies on contract and competition mechanism based on clear property rights, data utilization should be regarded as a more extensive cooperative undertaking. In the process of data generation, the activities of users provide important materials and power. In allocating the value created by the data utilization, everyone's interests should be included in a mutually beneficial structure so that the results of data value production are generally shared by the public.

Levying digital service tax is an important way to achieve this goal of value distribution. The main factor that hinders the use of data to benefit the general public is the digital gap: the differences in the information and communication technology in terms of ownership, skills, and application make some groups cannot share the benefits of the development of the digital economy, resulting in serious inequality. When the inequality caused by the digital divide exceeds a certain limit, the open opportunities for some people to use data for self-development will be reduced [26], thus thwarting the management goal of developing multiple values of data through the outbreak of creativity at the micro level. Therefore, it is necessary to make use of the economic benefits created by data as the goal of public sharing and settle on the group benefits in the most disadvantageous position of the digital society. There are two institutional paths towards this goal: either to ban the use of data that is not conducive to such groups from the source of value production, or to restore competitive neutrality through the levy of digital services tax. The shaping of the redistribution of benefits can better accommodate the social environment of multiple use of data. There is no moral justification for the choice of these two paths, but only a tradeoff. Even without considering the technical difficulties in practice, compared with the source prohibition, levying digital service tax can reduce the direct intervention of public authorities in the use of data and avoid excessive inhibition on the vitality of market subjects. Levying digital service tax is a better way to balance the moral requirements and policy objectives of value distribution.

6 Reconstruction of Data Protection - Utilization Legal System

The production of managing data value with the goal of innovation is the logical anchor of the "protection-utilization" legal system of data. The subject who practices in a playful way on the basis of autonomy is the source of innovative data value production. The intrinsic purpose of the field can provide normative guidance in the field of innovation. Tilt protection creative data utilization can provide direct institutional support for this process. Achieving universal benefits through redistribution can to some extent repair the inhibition of innovation caused by the dominant economic and social structure. In order to achieve this management goal, the specific system should create an environment that encourages innovative data value production. For this reason, this paper puts forward a kind of "data usefulness rules operating within the framework of extended personality rights and interests protection", which promotes the realization of the goal of innovative data value production management on the basis of implementing non-essentialism institutional logic.

6.1 Extended Framework of Personality Rights

In modern society, the subject is conceived as an individual who can choose freely, act independently and assume corresponding responsibilities. This concept of subject continues to the understanding and construction of data rights and interests in the contemporary legal system. In view of the fact that creative practice usually occurs in the process of direct encounter between man and nature, under the premise that the social environment leaves room for this encounter, the autonomy of the subject will naturally be accompanied by creative achievements. Therefore, the legal system with the autonomous subject as the core can promote the prosperity of human creative practice.

If the legal system in the digital age wants to achieve the same goal, it cannot stop at ensuring the autonomy of the subject. Computer algorithms have changed the medium of human practice in the digital age. The data presented by natural facts can be converted into code into calculation only if it is neatly organized under the carefully compiled category. Even if there is the slightest cognitive inconsistency between reality and code, prediction-oriented computing can bridge this gap through selective cognition and achieve overall unity [27]. Therefore, the artificial classification scheme and ontology structure have been embedded in the algorithm. When digital devices driven by computer algorithms infiltrate into all aspects of people's lives through interconnection, it makes code and software become the medium of cognition and action, thus cutting off people's encounter with the physical world. The technical architecture built by digital devices has even become the law of cyberspace, determining what people can and cannot do [28]. Only by creating new measures and possibilities can we continue to create new things and values. For regulating digital technology, the system of protecting personality rights and interests needs to shift from ensuring the autonomy of the subject to paying attention to its creative possibility.

This shift requires the "protection-utilization" legal system of data to expand the protection of personality rights and interests to supplement the shortcomings of traditional privacy protection and data self-determination. "expansion" mainly refers to the environment that gives the subject the guarantee of peace of life and the qualification of autonomy of the will to actually maintain these functions of the subject, especially to support creative practice. Subject is a completely abstract concept, which should be understood as the product of the interaction between creative self and social shaping factors. Only when this interaction can be realized without self-inhibition can creative value continue to emerge. Thus, it can be seen that the self-subject has media and social dependence. Because the algorithm logic only focuses on the statistical correlation between elements, but cannot be understood by ordinary individuals, digital media will actually affect the interaction between the self and the social environment. When people lack the understanding of digital architecture, it is difficult to bypass the restrictions imposed by the dominant value production model and creatively use data in a game way. When the digital architecture reduces the rich contingency under natural conditions because of the inherent artificial structure, the probability of people opening up new practice patterns due to the accident or accident of the environment is also reduced. Therefore, researchers should not only focus on people's data privacy and control, but also pay attention to the conditions needed to enable subjects to innovate data value production in a game and emerging way. And make the "protection-utilization" legal system of data pay attention to the design and operation practices that realize and maintain this condition. Therefore, the protection of data personality rights and interests should move from the traditional defensive and controlling traditional means to the "expanded personality rights protection" framework of constructing the creative practice conditions of the subject.

The "protection-utilization" legal system of data can realize the attention to the technical conditions of the "extended personality rights and interests protection" in two ways: (1) the governance of the embedded technical framework, and (2) the proposition of the availability of traditional media. Embedded supervision of digital architecture is the most direct way to achieve embedded governance. In addition to this kind of embedded governance, it can also give different subjects the right to advocate traditional technological media, hinder the technology architecture to fully cover people's daily life, reserve "breathing space" for self-development, and ensure the openness of space, information and cognition.

Although there are different possibilities in the path of the system implementation, the extended personality rights and interests protection of digital society serves the same purpose: to accommodate non-linear data value production. This framework does not directly adjust the behavior of data utilization, but to adjust the environment in which the data flow operates. The establishment of a good pipeline will not directly optimize the water quality but can improve the supply of water. Similarly, a healthy data transmission environment can make room for the subject's gameplay and self-development and establish the foundation for creative practice. Extended personality protection cannot replace data privacy protection and data autonomy based on informed consent rules, but to create conditions for the effectiveness of these traditional mechanisms. This framework which integrates the protection and use of data is more developed from the provisions of human dignity and freedom in the Constitution than from the personality right system in the Civil Code, so it takes precedence over the data usefulness system with the rights and interests of private law as the core. As the regulation of environmental or practical premise, expanding the protection of personality rights and interests should also have absolute priority over specific data utilization rules. Therefore, the extended protection of personality rights and interests has a lexicographic priority in the "protection-utilization" legal system of data: only on the premise of exhaustion of this framework can we enter into the tradeoff and application of data usefulness rules, or usefulness rules operate within the framework of extended personality rights and interests.

6.2 Data Utilization Framework with Goal of Promoting Value Production

The existing economic model takes the market as the core mechanism of resource allocation, and the "protection-utilization" legal system of data should ensure the effectiveness of the market mechanism in this field, especially in order to encourage the production of data value. Respect the productive input of data processors in data collection and processing. Under the premise of significant difficulties in the allocation of data property rights, data usefulness rules can meet the processing, control, research and development, license and even transfer of data ownership needs, protect and promote the production of data value. Most importantly, the "protection-utilization" legal system of data should accommodate developments and changes, and uniform rules for the use of data cannot be determined in advance. In view of the fact that the productive input of data processors is related to the meaning, determination and even planning of the value of their production data. In order to encourage the maximum development of data value, different usefulness rules should be set up according to this meaning to form a differential order pattern. The specific analysis is as follows:

6.2.1 Small Amounts of Productive Input Lacking a Clear Utilization Plan

The labor input in advance may change the moral status of the subject and enable it to obtain the legitimacy of monopolizing the use of data. At the same time, the establishment of exclusive ownership on data will increase the cost of data circulation, is not conducive to the free flow and access of information and has a negative impact on the production of data value. In order to optimize data value production, exclusive data rights should only protect productive inputs that demonstrate a determination to use the data or have a clear plan, rather than any effort made in the data collection process. Considering that the number of productive costs invested in collecting data can to some extent indicate the collector's determination to use the data, low productive inputs cannot provide any support for this subjective state. Therefore, the lack of low productive input in a clear utilization plan is not enough to constitute that the market subject has priority over the data controlled by himself. Other subjects do not need to obtain their own permission to use the data independently without violating other restrictions of laws and regulations (for example, the crime of trespassing into computer information systems). Even if the data controllers take corresponding confidentiality measures, they cannot regard this part of the data as trade secrets and resist the use of others. Although this regulation affects the supply side of the data industry due to the derogation of the value of data controlled by market entities specializing in data collection, it can still be effectively defended by reducing speculation in the data market and promoting the healthy flow of data from the source.

6.2.2 Moderate Productive Inputs or with a Plan that May Support Production of Desirable Value

The second case, the market subject actions show that they are willing to continue to invest in mining the value of data. The subjective will of the market subject can still be distinguished from the two aspects of labor achievement and the quantity of productive input. In terms of labor results, the market subject has left significant processing traces on the data set through selection and arrangement, although it has not yet reached the originality required by the compilation works. However, it may support the desirable plan for the market subject to continue to create the value of the relevant data. In terms of cost input, operators have obvious productive input, although they have not paid a lot of cost, but it is enough to show their determination to continue to mine the value of relevant data. These two aspects are usually closely related, and a certain scale of productive input can usually make the original data reflect the possibility of supporting certain desirable value combinations through collection and arrangement. In the case of meeting any kind of conditions, the market subject can be regarded as the willingness to continue to invest to create a desirable value combination, but it is not enough to guarantee to support this kind of data value production in a stable way. As a manager, arrangements should be made according to the development of the situation with a wait-and-see attitude.

In terms of system, this kind of "wait-and-see" is embodied in that it does not protect the original data collected by operators, but it makes the market subject in a specific legal position and can obtain some degree of monopoly through "further action". There are at least two situations of "further action": (1) data sets can be defended in the form of trade secrets under the premise of confidentiality measures; and (2) through further production inputs, develop data products or provide data services on the basis of relevant data sets. Under these conditions, market subjects enjoy competitive property rights and interests in respect of their products and services, which are protected by the general provisions of the Anti-unfair Competition Law. It can be seen that as the "further action" shows the determination and plan of the operator's production data value more and more clearly, the stronger the monopoly protection given by the legal system on the corresponding data controlled by the operator.

6.2.3 Large Amount of Productive Input or with a Specific Plan to Support Production of Desirable Value

In the third case, the actions of market players show that they have a strong desire to mine the value of data. Similarly, the subjective will of the market subject can be distinguished from the two aspects of labor achievement and the quantity of productive input. In terms of labor results, the data set processed by the operator can "obviously support a clear plan for the desired value combination". In terms of cost input, operators have invested a lot of labor and resources. Both aspects show that operators (compared with market players who are not willing to pay such investment) are likely to know how to use these data to create greater value and can support the production of relevant data value in a stable way. In this case, monopoly should be established through the legal system to create scarcity, and the data set controlled by the operator should be separated from the data Commons in a significant way. The most valuable system is the special rights rule of the EU Database Protection Directive, which gives database makers certain exclusive rights, which not only prohibit others from using and disseminating all or substantive parts of the database without permission, but also prohibit others from repeatedly and systematically using and disseminating the non-substantive contents of the database. The term of this exclusive right is fifteen years from the date of completion of the database. Considering that the purpose of establishing the exclusive right is to protect the realization of the operator's data value production goal, the period of monopoly protection should be commensurate with the reasonable cycle of data product or service innovation. It can be seen that the period of fifteen years is obviously too long, and the period of exclusive rights should be defined in the light of the stage of technological development.

During the period of this exclusive right, the development of events may present two situations. The first is that operators do not continue to make substantial use of the data and fail to further tap the value of the data. In this case, the special right status of the operator is lost with the end of the exclusive period. The second is that the operator makes the database or intelligent algorithm reach the standard of compiling works or patents by mining the value of the data, and the operator enjoys the copyright or patent right with reference to the intellectual property Law. The acquisition of copyright or patent does not lead to the loss of the exclusive right of the operator. The law should also encourage operators to continue to make productive inputs and create more value through continuous monopoly protection. There are at least two exceptions to the exclusive rights during this period of protection: (1) usefulness exhaustion: if an operator uses his exclusive right to data to seek a market monopoly, his exclusive use of the relevant data should be terminated;(2) Originality confrontation: other subjects obtain data controlled by exclusive rights subjects through web crawlers, but if their use of relevant data can produce multiple values and enrich people's social life style, it should be supported by the "protection-utilization" legal system of data. However, other subjects who cause damage to the competitive interests of exclusive owners while making use of the relevant data can request compensation, but this kind of compensation does not affect the legitimacy of data use behavior.

7 Conclusion

The new technological practice is constantly impacting the traditional ways of social life. In this context, emancipating the mind is inevitable to understand and reshape the current legal system and better respond to the reality. To emancipate the mind, there is great need to break the old ways of thinking and taking new perspectives.

By criticizing essentialism, the self-sustaining boring decoration of the old ways of thinking has fallen off, revealing the fundamental relationship between technology, legal system and human beings: (1) Technology is a survival strategy for people to releasing their own creative power in a cruel environment. (2) The key function of a Legal system is always to establish an appropriate social condition for creation. If human beings must be cruel and barbaric, it is necessary to return this barbaric creativity to mankind.

The anchor of data protection-utilization legal system is to release this creativity. On the basis of the application fulcrum of managing data value production, this article proposes a legal system of data usufructuary rules that operating within the framework of expanded personality rights protection.

Perhaps this kind of institutional conception is just the insignificant place, on which Marx or other great thinkers has set out. Every step forward is a major institutional project to reorganize a society that is splitting up in digital technology and redeem future of human beings from a gloomy prospect caused by restraining creativity.

References

1. Blackburn, S.: The Oxford Dictionary of Philosophy, p. 120. Oxford University Press (1994)
2. Danial, G.: Stroup: law and language: cardozo's jurisprudence and wittgenstein's philosophy. Valparaiso Univ. Law Rev. **18**, 331–371 (1984)
3. Cukier, K., Schönberger, M.V.: Big Data: a Revolution That Will Transform How We Live, Work, and Thank, p. 68. Mariner Books, Eamon Dolan/Houghton Mifflin Harcourt (2013)
4. Deakin, S., Markou, C.: Is Law Computable? Critical Perspectives on Law and Artificial Intelligence. Hart Publishing (2020)
5. Nietzsche, W.F.: Also, sprach Zarathustra, Manesse Biliothek, 144–145
6. Nietzsche, W.F.: Twilight of the Idols, p. 14. Oxford University Press (1998)
7. Marx, K.: DIE DEUTSCHE IDEOLOGIE, Hofenberg (2016)
8. Paul, O., Frankle, J.: Desirable inefficiency. Fla. Law Rev. **4**, 777–838 (2018)
9. Magnuson, W.: Blockchain Democracy: Technology, Law and the Rule of the Crowd. Cambridge University Press (2020)
10. Marx, K.: Economic and philosophic manuscripts of 1844. Dover Publications (2007)
11. Habermas, J.: Knowledge and Human Interests. Beacon Press (1972)
12. Bauman, Z.: Culture as Praxis, Sage Publications Ltd (1998)
13. Simon, H.A.: Administrative Behavior: a Study of Decision-Making Processes in Administrative Organizations. Free Press (1997)
14. Duxbury, N.: Normativity and technique in the legal philosophy of paul amselek, Arch. Philos. Law Soc. Philos. **75**(1), 104–111 (1989)
15. Razz, J.: The Practice of Value, p. 19. Oxford University Press (2003)
16. As the data protection - use part of the system logic, data value production value management is out of the limitations of essentialism, used to manage the concept: in the form of metaphor is ruled by the legal system to support or reject data using behavior decision-making, service to promote specific management objectives
17. Spengler, O.: Man and Technics. Greenwood Press, Translated by Charles Atkinson (1976)
18. Rouvroy, A.: The end(s) of critique: data behaviourism versus due process. In: Privacy, Due Process and the Computational Turn. Hildebrandt, M., De Vries, K. (eds.), pp. 143–168. Routledge/Taylor & Francis Group (2013)
19. BVerfGE65,1
20. Alexy, R.: A theory of constitutional right. Translated by Julian Rivers, p. 238. Oxford University Press (2002)
21. Strauss, D.: Philosophy: Discipline of the Disciplines, p. 560. Paideia press (2009)

22. Stahl, B.c., Andreou, A., et al.: Artificial intelligence for human flourishing. J. Bus. Res. **114**(4) (2020)
23. Wiener, N.: The Human Use Of Human Beings: Cybernetics and Society. Houghton Mifflin Co (1950)
24. Ashby, W.R.: Introduction to Cybernetics, Methuen Books (1976)
25. Webb, T.E.: Asylum and complexity: the vulnerable identity of law as a complex system. from jamie murray. In: Thomas, E., Webb, T.E., Wheatley, S., (eds.) Complexity Theory and Law. Routledge, p. 72 (2019)
26. Rawls, J.: A Theory of Justice. Belknap Press (1999)
27. Berry, D.: Critical Theory and the Digital, p. 196. Bloomsbury Publishing Inc (2014)
28. Lessig, L.: CODE Version 2.0, Basic Books (2006)

The Shifting Role of Government in Government Data Openness

Siqi Tao(✉), Yuan Xin, Zhao Li, and Minjie Hong

Law School, Jiangxi University of Finance and Economics, Nanchang 330013, Jiangxi, China
2201921727@stu.jxufe.edu.cn

Abstract. Data is the oil of the new digital era, and whoever gets the data gets the first chance of development. Government data is the data type with the most complete preservation and the largest scale in the data classification system. Open government data has become the focus of digital strategies in all countries. China has unique institutional advantages and is committed to forming a nationally integrated infrastructure institutional system and technical system: In terms of organizational construction, China has formed the National Data Bureau after institutional reform, which together with provincial data management bureaus forms an intensive and integrated digital government with central and local synergy, uniform standards and efficient operation. In terms of market construction, the central government policy encourages and local governments form a two-tier data market and authorized operation mechanism on an early and pilot basis, in order to give full play to the role of a competent government and an effective market, which is the centralized embodiment of national governance capacity modernization in the field of big data. Comparing with the EU government data space strategy, the Chinese government is committed to taking up the role of improving public welfare, bridging digital trust, and building a credible digital ecosystem. It can be found that the government's role changes from administrative regulator to cooperative organizer, from one-way bureaucratic governance to two-way interactive service, neither laissez-faire nor over-interference in the market formation process, and adopts a government-directed market-led policy approach to provide and guarantee the innovation environment of the digital economy.

Keywords: Data governance · Open government data · Promising government and the efficient market · Two-tier market of data element transaction

1 Introduction

The world economy has moved from industrial economy to digital economy era, and digital economy is the main economic form after agricultural economy and industrial economy, which is a new economic form with data resources as the key element, modern information network as the main carrier, convergence application of information and communication technology and digital transformation of all factors as the important driving force to promote fairness and efficiency more uniform. Under the development of digital economy, in order to strengthen the high-quality supply of data and innovate

S. Zhang et al. (Eds.): BigData 2023, LNCS 14203, pp. 80–94, 2023.
https://doi.org/10.1007/978-3-031-44725-9_6

the mechanism of data element development and utilization, data governance and data element market aim to build a credible circulation ecology of data and a market of sustainable utilization of data, respectively [1]. The EU and China have proposed different responses to this common challenge of the times, with the EU releasing a series of bills to facilitate the formation of the EU single market, and China taking advantage of its institutional advantages to jointly promote "Digital China" and "Data Element Market" through "central policy support and local early and pilot implementation". Among them, the open use of government data has become the focus of attention.

The main reasons for choosing open access to government data as a research perspective are, firstly, that government data originates from the public sector, contains public interests and is a public resource of society, which should be "taken from and used by"; secondly, compared with personal data, government data is less exposed to the risk of infringement of privacy and basic data rights, but due to historical reasons and popular beliefs, they face more challenges in Europe than in China; thirdly, due to the differences in national institutions and conditions, the EU and China's initiatives on open access to government data show the differences in the roles of the state and the market.

This paper will compare the initiatives of the EU and China in the open use of government data from the perspective of legislation and organization building, and analyze why and how China has formed a "centralized-decentralized-re-centralized" organizational structure with the unified guidance of the government and the cooperation of multiple entities in terms of history, national conditions, institutional mechanisms, and legal attributes of government data resources. The shift of the Chinese government's role in government data openness is consistent with the modernization of China's governance system and capacity, reflecting the advantages of centralized integration of government data resources under the public ownership system and the cooperation between the active government and the effective market under the socialist market economy. In anticipation of forming experiences that China and the EU can learn from each other in the open use of government data.

2 Open Data Act in Eu and China

2.1 Open Data Policy Under the National Strategy

The actions in EU's open data have been influenced by the Open Data Movement, with multifaceted content and sustained time. In the late 1980s, the European Union issued the Public Sector Information Directive (PSI), which focuses on improving collaboration between the public and private sectors in the information market. However, the needs of the private sector have not been fully addressed. It was not until 2002, European Action Plan established the EU Public Sector Information Development Framework proposal which has been realized the decentralized legislation and fragmented regulation in the EU, and hindered the development of information products and services in Europe [2]. This proposal was incorporated into the new PSI directive to coordinate openness and reuse of data, focusing mainly on government transparency and fair competition in the market, and providing institutional arrangements and unified rules such as Licensing System [3]. Since then, the PSI Directive has been reviewed and modified several times, with exponential growth in data availability and technologies.

The 2013 Amendment [the Directive (2013/37/EU)] officially mentioned the "right to reuse" of data, aiming at ensuring the social use of open data [4]. During 2007 and 2008, the EU balanced data regulation and data usage continuously, but it is clear that the EU's open data objectives still focused on improving government transparency and social governance. In addition, due to the fragmentation of the EU's legal framework for data protection and utilization, data protection law takes the main position while data market law and data governance law fail to follow up immediately. Furthermore, GDPR has established a high wall for enterprises data usage, with a fairly broad legal foundation that not only formalizes the process of data subjects exercising their rights, but also generates extremely high information costs and significant law enforcement costs. At the same time, the lack of general awareness between public and private sectors and technical difficulties, open data policy failed to help European countries occupy a leading position in the digital economy. Therefore, after 2017, the EU began actively developing a single market strategy aimed at exploring EU data flow issues, including access and transmission of machine-generated data, liability and security, data portability, data interoperability, etc., and committed to developing an overall policy framework that can promote the data value chain and improve the operating conditions of data intensive industries [5]. The overall objective of open data is to promote the availability of information, and improve the reusability of public sector information that not only contributes to the transparency and accountability of public administrations, but also brings additional attractions in terms of economic and technological innovation. On April 25 2018, the European Commission released the "Towards a Common European Data Space" communication accompanied with the package plan of "Building a European Data Economy", proposed a set of measures to form a common European data space, and aimed at developing new products and services based on data reuse and utilizing data to drive innovation for social benefits [6]. On February 19 2020, the European data strategy announced that the committee would invest in the European government data space in strategic economic sectors and public interest areas. The European data commons cover ten areas such as industry, healthcare, agriculture and finance.

The EU's historical tradition of human rights and concerns on digital capitalism have led to a strict model of data governance. Only a decade ago, data governance was considered to be the control and management of data. As the perception of the value of data has deepened, people tend to understand data governance as a cross-functional framework with specific rights, obligations, and formal systems for applications. Data governance research in the EU indicates that insufficient data sharing is a key challenge for organizational innovation, particularly in the field of public services. It is not just technical barriers that hinder data sharing, but also helps to understand the balance between the basic rights of data and the public interests under the development of the digital economy, and dispels the hidden worry about the uncontrolled development of data. The European Union's Digital Governance Act aims to establish new public-private partnerships and form trusted data intermediaries to bridge digital trust. At present, EU law provides abundant rights and value structures for data, but bottom-up individual actions often hinder the potential for improving the public interest, as personal data is public in nature and simple data protection is not the primary objective. "Public interests are not a choice between data protection and profits, but a struggle between different private value allocations and public collective benefits.

Compared to the open data policy in Europe, open data in China adopts a narrow definition. The subject of open data is only limited to the public sector. The concept of open data in Europe is relatively broad, covering all data within the public domain, regardless of whether it comes from the public or private sector, especially non-personal data collected and stored by the private sector such as internet technology companies, which also generates large scope of open data. Therefore, the subjects of data are not only governments, but also private-owned enterprises with public attributes. The open data in China mainly refers to the data formed by administrative agencies and public institutions performing public management and service in the course of performing public management duties and providing public services, which overlaps open data and government data in terms of scope. Some scholars believe that open data is the expression and integration of government data. However, there are differences between open data and government data in terms of political objectives. Open data focuses on the economic utilization of data, while government data focus on enhancing government transparency. In terms of the actual subjects of release, these two are the same batch of data. Some scholars consider that open data is relative to personal and enterprise data in classification. Open data is a fair, reasonable and non-discriminatory social information resource that contains the public interests of society [7]. Therefore releasing open data is the duty of government in placing "service-oriented" function and reducing the cost of information [8]. From the perspective of policy and legislation, the expression of open data is inconsistent at the national level. In 2015, the expression in the "Platform for Action to Promote Big data Development" is, "steadily promote the opening of open data". [9] in 2017, the expression "public information resources" and "government information resources" is adopted. The Big data development action in 2015 has linked with the existing government informatization reform. The early work of open data in the political and legal fields focuses on building an open and transparent government, taking the opening of public information resources as an opportunity to provide a platform for the centralization of social resources, rather than directly transforming government information resources into available economic resources. At this stage, it is necessary to distinguish between the objectives and effects of open data. The "service-oriented" government requires government informatization to improve the efficiency and effectiveness. It aims at the transformation and upgrading of governance, reduces the additional social information costs in the process of upgrading, and reflects the positive Externality of governance objectives in terms of economic effects.

Thus, there is an inherent governance logic and economic logic embedded in the topic of open access to government data, and the transfer of the two. The exchange is phased and ephemeral. How can government data be governed so that what was originally a governmental resource is transformed into an economic resource that has economic value and can be opened up and reused? Which organization (who) will build this infrastructure, and in what way the costs will be borne, depends on how the economic value of government data and its legal attributes are perceived. The answer to the question of who owns government data determines who, in what form and for what purpose it is developed and governed, how the costs and benefits are allocated and how the risks are bore in the process of opening it up for use.

2.2 The Logical Transformation in China's Government Data Openness

In the transformation from government data as government resources to government data as a factor of production, we face a series of problems.

Firstly, who is responsible for the collection and management of government data? It is the data collected and stored by government departments in the process of performing public management and service functions, which means that the collection and storage of data by government departments is not intended to become a factor of production, but only as a supplement to market data resources. More importantly, The development of digital economy depends on the concerted participation of the state and society as the builders of data infrastructure to achieve "co-construction, co-governance and sharing". Specifically speaking, as a government resource, government data need to go through the economic process of production resource and capitalization of assets, and be reviewed and processed in line with data security. Only the infrastructure established by public-private cooperation can be realized. In terms of specific systems, the government's technical and management capabilities, the cost of government assets and how to provide effective incentives for the government should be considered. It requires the government to assume governance and economic responsibilities in the data elements, calling for the transformation of a promising government.

Secondly, who will develop and use government data? Since 2018, in response to the market-oriented policy of data elements, data exchanges and data trading centers have mushroomed in China's regional areas. However, according to statistics, most of the data exchanges have not achieved their intended purpose. The data exchanges in Shanghai, Guangzhou and Shenzhen have performed well. The reasons are not only the technical limitations, but also the practical differences in the development of local data industry, the absence of institutional incentives and trading rules, as well as the immaturity of trading subjects and trading modes. The overall expectation is to make up for the market.

Thirdly, how to share the costs and benefits of development and utilization? Government data utilization often faces the criticism of "tragedy of the Commons" and "free rider". It is especially sensitive on how to confirm the property right of government data and how to balance the cost-benefit mechanism of the government as the builder of public land. Some scholars believe that government data belongs to the government [10]. Some scholars believe that government data as a public property, should weaken the emphasis on ownership and shift to the protection of the right to use, avoid the exclusive use of specific subjects, [11] promote the open sharing of government data resources, and ensure the maximum value of government data resources in the social scope through public use and collective actions.

2.3 Prerequisite for Realizing the Value of Government Data

Currently, the open use of government data has entered a new stage dominated by economic logic. Data utilization refers to the process of processing data through steps such as cleaning, analysis, integration, correlation, and visualization to generate data products and services. It is an important step in realizing the value-added potential of open data [12]. Most studies indicate that data utilization exhibits industry-specific characteristics and inequalities in various domains. Individuals, being in a weak technological position, face challenges in data protection and utilization, requiring legal protection to address these issues [13]. The failure of an empowerment system based on "informed consent," the hidden risks of harm, and the ambiguity of remedies call for national regulatory measures to fulfill data protection obligations [14]. Data utilization and governance have risen from technical issues to national responsibilities.

Data utilization, as a neutral technical process, is value-neutral in itself. However, it has entered the realm of data law due to several reasons:

First, data utilization is a critical link in the generation of data value, transitioning data from a production factor to data products and services. Data utilization connects data generation and data processing, representing the technical expression of data commodification. Ownership issues between data holders and data processors arise in data utilization, as well as value distribution issues during the initial stages of data.

Second, data utilization is a multi-stage technical processing activity that requires comprehensive management and supervision throughout the process. This includes preventing potential risks of data leakage and aggregation, as well as the process of reanalyzing and integrating data after cleaning to generate and discover new social relationships. Data aggregation, recombination, matching, and distribution not only reflect the original content of the data but also generate new data and discover new associations, which are the main clues for economic value generation. In other words, this includes but is not limited to the requirement for anonymizing original data involving personal information and ensuring that the algorithmic framework used for integration and analysis is objective, neutral, and free from bias and discrimination.

Third, data utilization requires technical capabilities and corresponding technological literacy. In other words, groups without technical capabilities and literacy are excluded from the scope of data utilization from the beginning. Apart from professional technical personnel, the vast majority of data subjects can only exist as data consumers. They generate data and are subjected to instrumental logic in data environments for calculation and consumption. They have no control over the use of data and cannot prevent data misuse. This has led to discussions on data ethics and criticisms of data capitalism. The development of the Internet and emerging technologies has intensified the "surveillance-industrial complex". Western scholars have expressed concerns that the collection, storage, control, and analysis of big data are conducted under the backdrop of political and economic interests. The purpose is to exercise economic and political control over individuals and treat them as targets. The convergence of surveillance capitalism and surveillance by the state is a perfect match between Big Brother's power and the interests of capitalists. The ownership and control of data have led to a "data divide".

Fourth, data utilization exhibits two trends: the privatization of private platforms and the publicization of public service platforms. In fact, Internet platforms have gradually gained organizational power beyond geographical boundaries. They shape digital markets and build corresponding infrastructures, forming stable production orders. They respond to national policies and implement governance goals in economic development. At the same time, governments have encouraged the landing of new production models by relaxing market access and weakening policies on operator concentration. They position platform enterprises as governance entities rather than objects [16]. In terms of the organization model for public services, intelligent platforms and offline windows together bear most of the public functions. They enhance on-site law enforcement capabilities, improve the efficiency of public services through "data mobility" and "policy finding people". For example, the push and notifications of service platforms such as smart cities, smart governments, and smart healthcare, as well as the assistance of health code mini-programs during the pandemic for public health management. This platform-oriented model of government services places higher demands on the government's informatization capabilities. Firstly, they need to have technological research and development capabilities to provide software and hardware products that meet the basic needs of government services. Secondly, they need the capability to operate and maintain public service platforms, which involves interconnecting data between platforms and the reintegration and supervisory management of governmental responsibilities.

In summary, data utilization is not simply a neutral technical process; it also carries multiple interests and diverse values. It represents the "greatest common denominator" of personal data protection, national data security, and socio-economic value. The sources of government data, the entities involved in its utilization, and the value it drives exhibit a complex fusion of public and private interests. To compensate for the limited action capacity of governments and their departments in public services, practices often adopt the form of entrusting specific companies rather than promoting fair public utilization of the infrastructure construction of government data utilization [17]. In addition, in government data trading platforms and markets developed in various regions, there is a gap in connecting the supply and demand ends of data. The absence of market systems and mechanisms hinders the development of a thriving government data market.

3 Bridging Government Roles in Data Governance and Data Utilization

3.1 The Dilemma of Organizational Structure in Government

Although 208 local data open portals have been launched nationwide as of October 2022, in terms of overall social perception, the open data work still suffers from low data value, slow data update, and poor data quality. The root of the problem lies in two points: "can't" and "won't". On the one hand, it is "can't", the data opening work involves a high degree of technical expertise and needs to deal with personalized needs directly to the data users, so traditional public departments lack the corresponding technical and service capabilities to do a good job of data opening. On the other hand, it is "unwilling", that is, data opening requires free data public welfare, and in order to meet social needs, to ensure data security and quality, it also requires financial, human and material investment [18].

In the internal organization of the government, the traditional hierarchical system has encountered conflicts of division of labor and contradictions of authority and responsibility in the transformation of the network logic of the platform government [19]. The traditional hierarchy system changes in administrative organization structure, communication form, and internal control means [20]. The original individual instructions from superiors to subordinates are replaced by user needs and service structure in the logic of networked flattening; the automated of the administration and procedural justice also face ethical questioning; the digital administrative structure cannot meet the demand of clear power and responsibility distribution, and the traditional hierarchical structure administration has clear division of responsibilities and fulfillment. However, in the networked automatic administration, the complex administrative needs may fall into the rule of law trap of "mixed administration" and "unclear responsibility".

In terms of governmental function transformation, the administrative system undergoes a shift from manager-centric to user- centric, with a service mindset replacing the previous managerial mindset. The standardized, sectorial, efficiency-focused traditional hierarchy system moves to an electronic government based on a coordinated and interactive network (centralized decision-making and administrative sectioning). The platform government requires breaking the vertical hierarchical reporting organizational structure, and the original logic of doing business requires citizens to adapt to the management-oriented organizational design, and to run "horizontally and vertically" according to the window, department and hierarchical authority, and the internal sharing of government data can help "only one run", "one network to do" and "one network to manage". In the use of government data, data sharing has a smaller scope than data opening, and the demand for data processing is clearer, but the reason why sharing has not yet "opened the last mile" lies in the lack of data barriers and coordination mechanisms in the sector. The data barriers have both technical and organizational flaws. Technically, there are gaps in the data capabilities of various departments at different levels, and the grassroots departments even lack professional and technical personnel; due to different industry sectors, departmental levels, security risks and other factors, data types, directories and formats do not necessarily correspond to each other, and the standardized and secure governance of departmental data lacks professional and unified guidance; in the organizational system, there is a lack of unified supervision and control of data quality, and the timeliness, accuracy, interoperability, and feasibility of the data are not always guaranteed. In the organizational system, there is no unified supervision and control of data quality, and the timeliness, accuracy, interoperability, and machine readability of data cannot be guaranteed.

3.2 Organizational Restructuring for Government Data Openness

Karl Marx points out that every deep penetration of the new technological revolution will inevitably lead to changes in the social structure and social formation. As a kind of subversive information technology revolution, big data has profoundly changed the production and living conditions of the society and reshaped the governance ecology of the country. As noted above, big data poses new challenges to hierarchical organizational structures while improving governance instruments. Government departments are faced with not only the information that digital management can collect, organize and grasp, but also the continuous generation, update and decentralized storage of big data in the way of social needs, challenge the country's ability to govern. The influence of the state to the society is not one-way, the society has the reverse effect to the state. In stable societies, the capacities and functions of the state are relatively fixed, the state and society are nested within each other, and the capacities of the state are consistent with the needs of society, but this consistency is not constant, following up according to the development of the society to learn. Faced with the fluid and uncertain society, the process of national response is essentially the acquisition of national capacity, through continuous learning and absorption of new technology ("Administrative absorption technology") to achieve the improvement of governance capacity.

The open utilization of government data is only a microcosm of the society which is constructed by "Data". The reason why government data is special is that it is a factor of production for the development of digital economy, it is also a resource of governance in the data age, which means that the state's mastery, use and transformation of this factor of production is particularly important. On the one hand, the digital technology is embedded in the national hierarchical system. The technology becomes the means of governance. The technological logic reinvents the organizational process, and the national digital capacity needs to rely on the technical support provided by the private sector to strengthen. At the same time, it shows that digital technology enables subjects with technological advantages to acquire de facto power and governance capabilities, and social subjects shift from monism to pluralism [21]. On the other hand, countries need to create a stable order through institutional design for fair access and widespread use of data. Government data moves from idle to available resources, working in ways that ease Information asymmetry and reduce transaction costs in the production of value. Property regimes allocate resources in a way that provides a set of rules for combining people and means of production, but exclusive property rights are not optimal for the data. Since data can be copied and transmitted indefinitely, the common use of government data by most people does not detract from the value of the use of that data by others, nor does it result in the scarcity of data resources. On the contrary, it can promote the use of government data in different dimensions and increase the differences and competitiveness of market products. "Absolute exclusive data property rights not only fail to protect users' privacy, but also hinder the maximum exploitation of data elements, so no one should have exclusive property rights to data elements themselves" [22]. Data access is more important than data property rights because of the non-competitive use of data. Some scholars further point out that the emphasis on "National ownership" of data resources may strengthen the inter-departmental ownership of data from the internal, stimulating the motivation of public institutions to operate and profit impulse. Thus, the modernization of state capacity does not require the establishment of an "All-powerful" government. On the contrary, in the face of the complexity of social governance, the capacity of the government should be limited to a certain range, delimit the boundaries of the capacity, and strengthen the cooperation and synergy with social organizations. The role of government has changed from administrative supervisor to cooperative organizer, from one-way bureaucratic governance to two-way interactive service. In the process of market formation, the government adopts the policy of "Government guidance and market leading" to provide and protect the innovation environment of digital economy.

In order to better connect the work of government informatization and data resourcefulness, the state has formed a new agency to centralize big data management functions. In 2023, the State Council submitted for consideration the institutional reform program decided to establish the National Data Bureau, which will be managed by the National Development and Reform Commission, and transferred the work functions of the Office of the Central Cyberspace Affairs Commission and Informatization related to coordinating and promoting the integration, development and sharing of national information resources and promoting the informatization of social governance to the Bureau. The National Development and Reform Commission in relation to the coordination of the development of the digital economy, the promotion of data infrastructure and data elements of the basic system of responsibility together into. This has enabled the simultaneous promotion of sectoral unification at three levels: data industry, data infrastructure and data trading system at the central level with a specialized and integrated departmental setup. In fact, before the establishment of the National Data Bureau, localities have begun to set up big data bureaus to explore the connotation and extension of big data administrative functions, divide departmental responsibilities, and build a digital government system that is jointly built and shared by multiple subjects. The adjustment of the state in the administrative organization structure generally shows the trend of the government from focusing on industry management and government management functions to the comprehensive governance of big data.

By contrast, the governance structure of the Common European Data Spaces consists of "multiple data spaces plus a central authority". The Data Governance Act proposes a central body, The "European Data Innovation Board", as the strategic level authority, and the Data Innovation Board and Data Exchange Board, the strategic and operational level governance bodies respectively, with a central body composed of representatives from all sectors of the data space. Provides oversight and advice to ensure democratic decision-making and a level playing field. Correspondingly, the governance body for each industry data space is aligned with the overall governance body. A central governance body establishes the enabling framework, including rules and standards, and oversees all matters related to data space interoperability - the de facto "soft infrastructure". In the data space, individuals, governments and enterprises are the "users" (called "joiners") and "participants" of the data space, providing identity, authorization registration, function testing, participant authentication, security testing, etc. through data intermediary service providers (called "authenticators"). A broad and diverse set of certified parties are subject to the "soft infrastructure" framework developed by the Data Exchange Board, with data owners/data providers, data consumers, metadata proxy service providers, clearinghouses, application store providers, and others as governors of the International Data Space (IDS).

4 Government and Market in Trusted Digital Ecosystem

4.1 The Cooperation Between the Promising Government and the Efficient Market

As mentioned above, in the digital society, the relationship between the government and the market has become closer, and the social organizations outside the government and the private sector have gradually become important supplements and collaborators of government subjects. In the formation of the government data ecosystem, the exercise of government functions depends on the screening and subcontracting of the market mechanism, and the large platform and large enterprises in the market have become the collaborators of social governance. The use of government data realized in the way of public-private cooperation is in line with the law of the formation of the digital market, which can effectively make up for the limitations of the government itself and allow professional people to do professional things, so as to realize the mutual benefit of digital government and digital economy. The main responsibility of the government is reflected in the mobilization and integration of social subjects and resources, the provision of fair institutional guarantee and the preliminary construction of infrastructure, and its function should be located in ensuring that multiple subjects obtain and use government data under fair and reasonable conditions. As for the transaction details and transaction structure of data applications, which are not sought by the government but promoted, discovered and improved by market entities in the process of mobility, the government should strive to guide the endogenous growth of the market and the endogenous vitality of social innovation, promote the sharing and opening of data between departments and regions, protect the property rights and legitimate interests of all kinds of market subjects in accordance with the law, and create a stable, fair, transparent and predictable business environment.

There is a promising government that can provide social order guarantee, maintain the development environment and provide the protection of citizens' rights. The government has the responsibility to strengthen supervision through institutional innovation, while applying technology to ensure data standardization, anonymization and desensitization, complete data resource utilization, and avoid externalities such as privacy disclosure, information overload, data pollution and so on. Efficient market means to give full play to the decisive role of market mechanism in resource allocation and let market economy laws such as the law of value, the law of competition and the law of supply and demand play a role. In the establishment of the government data factor market, the allocation of data resource elements is constantly improved in the process of change, and the "Pareto Optimality" is realized to match the most suitable market structure.

There are stages for the organic combination of the promising government and the effective market, and the status of the government and the market is different in different stages, which is reflected in the gradual weakening of government guidance with the maturity of the market. From the point of view of the optimization of the market structure, in the first stage, when the market is still in its infancy and the industrial scale and system are not yet perfect, it is appropriate to adopt the path of "government guidance", that is, the government provides the market plan and specifically implement and control the process of market reform and change. On the one hand, it can effectively attract industrial agglomeration, on the other hand, it can standardize the market trading order and form a good market ecology. In the second stage, with the maturity of the market and the emergence of government-led disadvantages, we should choose the path of the combination of "government guidance" and "market follow-up". The government should guide the market through publicity and training to induce the motivation of market optimization. The market follows up. In the third stage, with the continuous increase and improvement of the market scale and structure of data elements, the optimization path of "market-oriented" can be adopted, and the government plays the role of "night watchman".

4.2 Construct the Two-Tier Market of Data Element Transaction

In order to give full play to the value of data, compliance governance and value realization of data elements are the premise and goal of each other, and should not be neglected. The governance with the participation of all kinds of subjects should be the collaborative governance guided by the government and led by the market. The underlying institutional logic of local data transaction exploration reveals the feasible guidelines for the opening of government data – first, to build a primary market with data sources on the basis of data operation and management institutions, collect and govern government data to control data security risks from data sources, and build a credible and efficient data trading market environment. Second, incubate and cultivate multiple subjects in the process of data transactions, such as data intermediaries, data processors, etc., to provide professional market power for the formation and transaction of data products and services, and promote the formation and development of the secondary market with data users. Third, the formation and supervision of operators are involved in the formation of the two-tier market, and the authorized operation mode can be adopted to solve the executable problem of data openness. The authorized operation of government data does not create a new open form, it provides how to make the current open form of data executable, focusing on the "subject of action", that is, "who will do it". The key is that the current data authorities authorize third parties through contracts and other means, allowing third parties to intervene with professional strength to carry out data flow. "Authorization" is the transitional product of the establishment and development of the data element market, which solves the preliminary "trust" problem of the data element market.

China's data factor trading market is still in the period of exploration and development. As the primary market, data trading center needs to strengthen the data transaction protection mechanism to avoid the risk of data security. Data enterprises should be encouraged to use technologies such as blockchain, distributed computing and privacy computing to participate in the development of data security products and technical solutions oriented to the primary market of data elements, and to create a credible environment for data security transactions by technical means. Strengthen the protection of data applications and solve the problems of uncontrollable use of data and easy disclosure of private data. For example, the Beijing International big data Exchange deconstructs the data elements into visible "specific information" and available "computational value", confirming, depositing and trading the "computational value", realizing the "invisibility, controllability and metrology" of data circulation, and providing a credible data fusion computing environment for both data supply and demand. Shanghai data Exchange puts forward the principle of "unqualified non-listing, no scene no trading". Zhejiang big data Trading Center launched the big data power confirmation platform, using the open source big data distributed computing framework, original "data burning after use" and other technical solutions.

In the data secondary market, multiple data subjects such as "digital quotient" and data brokers are encouraged to enter the circulation market, multiple data product portfolios with the combination of data, algorithms and computing power are encouraged to expand the form of data assets. For example, Beijing International big data Exchange pioneered a "digital trading contract" model based on blockchain, which broke through the traditional primary model of single data trading and developed into a combined trading model covering data, algorithms and computing power. It expands the value realization scope of data resources, and turns algorithms, computing power and integrated service applications into digital assets for trading. Shanghai data Exchange pioneered the "digital quotient" model to guide multiple subjects to increase data supply.

5 Conclusions and Future Work

Open data is driven by four dimensions: (1) In terms of culture and rights, knowledge belongs to all humanity, and knowledge acquisition is a fundamental human right as a typical concept. (2) In the political and legal fields, open data implies a more open government, with stronger participation inclusiveness, transparency, and accountability. (3) The typical expression in the field of market and economy is that data is oil and data is infrastructure. (4) In the field of technology and ethics, data provides the foundation for ethical development, application of AI and automated algorithmic decision-making. With the development of AI and large-scale algorithms, the structure and volume of data required for AI application iteration place new demands for the release of open data. The capabilities of machine image recognition and speech recognition open up the scope for expanding the sources of data collection and the ability of large collections of unstructured information, and generate datasets for machine learning. In other words, the expansion of AI capabilities changes the requirements and scope of machine readability, and is no longer limited to supporting machine learning with pre-existing raw materials. The development of AI expects a more real-time, complete, and diverse data pool, which means higher requirements for data collection, standardization, and sharing methods. Some scholars believe that the release of open data creates opportunities for civil society to participate in shaping open data infrastructure in certain aspects. However, some scholars doubted that only by opening data would not generate value. The key step to discover open data value is to find appropriate business models. It is not sustainable to rely solely on the call and vision of open data to encourage enterprises to open up data. Legitimate motives and decisions through open data policy may not take into account the full complexity of the open data business model. Motivation and beliefs need to be aligned with investment in order to create value.

Thus, the opening of government data has put forward many specific requirements for government departments, but the reality is that the opening is still insufficient compared with the demand for utilization. Moreover, full openness does not necessarily lead to full utilization. Because the basic logic and value pursuit of the traditional government and the market are not consistent. The effectiveness of data utilization is closely related to data governance. In order to reduce or eliminate the negative externalities of government data opening as much as possible, it is necessary to control the risks in the process of data utilization and regulate the behaviors in data utilization with the help of technology, system and industry regulations. Changing the structure of government organizations to adapt to digital transformation and creating a larger and more large-scale government data pool is the first step, and the initial stage of controlling data risk at the source. The common problem facing all countries is to bridge the gap between openness and utilization, build digital trust supported by institutions and mechanisms, and create a stable order for fair access and extensive use of data. This requires the government to be the organizer of cooperation and the provider of public services. China attempts to provide a feasible solution by establishing a Two-tier market and authorized operation mechanism. As the provider of a centralized platform, the government brings multiple social entities into the data ecosystem to jointly shape, jointly manage and jointly enjoy the overall welfare brought by data, which is a decentralized structure. How to shape the data ecosystem will be a larger but more nuanced issue that needs to be discussed separately.

This paper is the research result of the National Social Science Foundation project "Research on legal Governance system and Legislative change in Digital Society" (20&ZD178).

References

1. Lessig, L.: Code version 2.0 (2006). http://codev2.cc. Accessed June 2023
2. European Commission, Directorate-General for Communications Networks, Content and Technology, Linz, F., Vries, M., Barbero, M., et al., Study to support the review of Directive 2003/98/EC on the re-use of public sector information – Executive summary, Publications Office (2018). https://data.europa.eu/doi/10.2759/931204
3. Burwell, S.M., et al.: Open Data Policy — Managing Information as an Asset (2013). https://digital.gov/resources/open-data-policy-m-13-13/. Accessed July 2023
4. Publications Office of the European Union, Carrara, W., Radu, C., Vollers, H., Open data maturity in Europe 2017: Open data for a European data economy, Publications Office of the European Union (2020). https://data.europa.eu/doi/10.2830/918627
5. European Commission: (2015) Communication from the Commission to the European Parliament, the Council, the European Economic and Social Committee and The Committee of The Regions 'A Digital Single Market Strategy for Europe' SWD (2015)
6. European Commission: (2018e) Communication from the Commission to the European Parliament, the Council, the European Economic and Social Committee and The Committee of the Regions "Towards a common European data space" COM (2018)
7. Zhang, T.: Hiding wisdom for the people: legal basis and norm remodeling of open government data. E-Government **200**(08), 75–90 (2019). (in Chinese)
8. Gao, Z.: The value positioning and implementation path of government data open system. Digit. Libr. Forum **188**(01), 27–34 (2020). (in Chinese)
9. State Council: Action Plan to Promote Big Data Development (2015). https://www.gov.cn/gongbao/content/2015/content_2929345.htm. Accessed June 2023
10. Zhao, J.: The rationality and legal significance of government data belonging to the government. J. Henan Univ. Econ. Law **36**(01), 13–22 (2021). (in Chinese)
11. Qi, Y.: Construction of rules for the use of government data resources as public property. Adm. Law Rev. **129**(05), 138–147 (2021). (in Chinese)
12. Gao, F.: Rights allocation of data holders - legal implementation of structural separation of data property rights. J. Comp. Law. (in Chinese). http://kns.cnki.net/kcms/detail/11.3171.D.20230526.1419.004.html
13. Ding, X.: On the legal basis and institutional framework of oblique protection law. Peking Univ. Law J. **79**(07), 112–114 (2022). (in Chinese)
14. Wang, X.: National protection obligations and implementation of personal information. China Legal Sci. **219**(01), 145–166 (2021). (in Chinese)
15. Sai, S., et al.: Research on digital capitalism by foreign Marxist scholars. World Soc. Stud. **8**(03), 98–108+112 (2023)

16. Hu, L.: Two types of data order and legal responses. J. Soc. Sci. **512**(04), 181–192 (2023)
17. Wang, X., et al.: Establish a government data opening system based on the right to fair use. Soc. Sci. Dig. **76**(04), 12–14 (2022)
18. Jing, J.: Organizational approach to administrative digitalization. J. Nanjing Univ. **60**(01), 116–123+162–163 (2023)
19. Zhang, X., et al.: The digital shadow of public administration: ethical conflicts in building a digital government. J. Public Adm. **15**(05), 164–181+200 (2022)
20. Chen, J.: The basic logic of legitimacy certification of automated administration. Legal Forum **38**(03), 50–62 (2023)
21. Zhang, L.: Public-private partnerships in national digital capacity building. Law Polit. Sci. **01**, 46–55 (2022)
22. Tang, J.: Economic analysis of data property rights. Soc. Sci. J. (01), 98–106+209 (2021)
23. Chu, J., et al.: Research on government responsibility of government data openness. J. Mod. Inf. **39**(10), 127–135 (2019)
24. Wang, X., et al.: The right to fair use: the right basis for the construction of government data open system. ECUPL J. **25**(02), 59–72 (2022)

How Do We Talk and Feel About COVID-19? Sentiment Analysis of Twitter Topics

Carmela Comito(✉)

Institute for High Performance Computing and Networking (ICAR), National Research Council (CNR), Rende, Italy
carmela.comito@icar.cnr.it

Abstract. COVID-19 is the subject of intense and widespread discussion on social media that de facto became one of the main means for people to get and share news about the pandemic. Social media discussions may influence public opinions and could also disseminate panic and misinformation during crisis events like COVID-19 outbreak. In this context, it is crucial to detect the topics being discussed on social media and understand people perceptions, opinions and feelings by analyzing the sentiments of users towards those topics. Accordingly, this paper proposes a topic-aware sentiment analysis model. The main element of novelty of the model is that it computes the sentiment at topic level by applying a multi-label classification approach on top of the online clustering detection of the topics. The approach has been validated over a real dataset of tweets about COVID-19 in US. Results highlight that the proposed method correctly identifies the sentiment of the relevant topics like the preventive measures adopted or the curative means used. The evaluation demonstrated that the proposed sentiment classification algorithm showed higher performance compared to traditional methodologies.

Keywords: COVID-19 · Social Media Data · Topic Modeling · Sentiment Analysis · Twitter

1 Introduction

The worldwide and fast diffusion of COVID-19 pandemic resulted also in a massive rapid spreading of pandemics information on a global scale. Online social media, such as Twitter and Facebook, have been and are still extensively used to report and share news, events, updates, sentiment about COVID-19. As a consequence, a large amount of data about the pandemic is spreading in real-time over social networks. This large amount of data can be exploited by medical and government institutions to understand people dynamics and adopt preventive and corrective measures to confine COVID-19 effects. Accordingly, a strong need of efficient analytics methods and tools to comprehend such data arisen.

S. Zhang et al. (Eds.): BigData 2023, LNCS 14203, pp. 95–107, 2023.
https://doi.org/10.1007/978-3-031-44725-9_7

This article focuses on the analysis of the conversations on social media, specifically Twitter, with respect to COVID-19. During the COVID-19 pandemic, people have been using social media to share not only information but also daily activities and thoughts, including feelings about their personal situation, health status, suggestions on how to behave and treat the symptoms. Such information may supply large-scale behavioral insights about people reaction to the pandemic. However, it is not easy to identify and extract meaningful and useful information from the noisy and short text characterizing social media posts.

In recent years, sentiment analysis has been proposed to analyse social media to identify and study affective states and subjective information of people. It is a branch of affective computing that exploits natural language processing, text analysis, computational linguistics, and biometrics in order to classify a textual content as either positive, negative or neutral. With the COVID-19 emergency, understanding the sentiments and the emotions of the people is extremely important. Preventive measures like lockdown, social distancing, quarantine might affect people's life introducing psychological implications and even compromising mental health.

Traditional sentiment analysis techniques aim at classifying the overall sentiment of a text without considering other key features of the content such as towards which topic or aspect the sentiment is referred to. The objective of this work is to identify the most significant topics that are useful to understand the COVID-19 pandemic situation and then assess users opinion and feeling toward those topics. Specifically, in the paper is proposed a topic-level sentiment analysis model in which topics are extracted from the tweets and then sentiment analysis with reference to the extracted topics is performed. The proposed approach merges methodologies from two research areas, i.e., topic detection of social media streaming data and sentiment analysis. To this aim first we extract and analyze all topics related to the COVID-19 by clustering bursty textual features of tweets. Once the topic is detected, it is analyzed by a classifier for sentiment detection by computing the polarity as positive or negative or neutral. In this way, topic-based sentiment analysis aims to give sentiments towards the topics by looking at the tweets in the topics and a set of key textual features (e.g., hashtags) in a give temporal period. It is worth highlighting the importance of the temporal variable since the sentiment for a given topic could vary over time as it depends on the bursty textual features that are usually bursty for a limited time horizon.

A real-world dataset of COVID-19 tweets has been used to evaluate the proposed approach. Results shown that the method is effective in detecting topics of discussions related to COVID-19 and to compute the sentiment of the topics, outperforming traditional approaches. Public sentiments related to COVID-19 spreading and extracted from the tweets, show people opinion and feelings towards the detected topics that actually represent the main issues the world are facing after COVID-19 outbreak.

The rest of the paper is organized as follows. Section 2 overviews related work. Section 3 presents the proposed approach: Sect. 3.1 formulates the problem, introducing the key aspects of the approach, while Sect. 3.2 presents the topic sentiment detection algorithm. The results of the evaluation performed over the real-world dataset of tweets are shown in Sect. 4. Section 5 concludes the paper.

2 Related Work

Starting from the first days of COVID-19 outbreaks several research activities exploiting social media data to address COVID-19 issues have been launched. Some of such activities focused on the study of human behavior and reactions to COVID-19 [1,9,12,13]. Conspiracy detection is another topic widely explored [6,14] as well as misinformation propagation about COVID-19 [11,16].

COVID-19 topic detection is another well studied research line. By the best of our knowledge the majority of the proposals in literature aiming at detecting COVID-19 topics on social networks exploits the *Latent Dirichlet Allocation (LDA)* model [2], like the following references [1,7,10,15,16]. *LDA* is a topic model where words and documents are linked by means of latent topics. The model computes for each textual content a probability distribution over topics, which are distributions over words. The main relevant drawback of LDA is that the number of topics should be fixed in advanced and the data structure and size has also to be fixed in advanced. To overcome the limitations of current state-of-the-art, in [3] has been proposed an approach, presented in [4], for COVID-19 topic detection in Twitter that combines peak detection and clustering techniques. Space-time features are extracted from the tweets and modeled as time series. After that, peaks are detected from the time series, and peaks of textual features are clustered based on the co-occurrence in the tweets. Each cluster obtained is then associated to a topic.

3 Topic-Specific Sentiment Classifier

The section describes the proposed topic-specific sentiment classification model by first formulating the addressed problem and then introducing the algorithm and its main components.

3.1 Problem Formulation

In this paper, we investigate how discovering the topics discussed in a set of tweets can be used to improve the sentiment classification of Twitter users.

A textual content of a tweet usually refers to one or more topics. Therefore, grouping the tweets on the basis of the topic of discussion expresses better the content of the discussions taking place on social media and enhances the sentiment extraction process. Motivated by these challenges, this paper proposes an online learning model for sentiment analysis of topics. Specifically, the aim is to detect the topics in an online manner to deal with streaming data and then perform topic-level sentiment analysis. For detection of topics from streaming data, an incremental clustering algorithm has been proposed. Standard approaches to topic modeling like LDA requires that the number of topics must be set beforehand as well as the data structure and dimension. Clearly, in dynamic and evolving settings like social media this it is not feasible. To overcome this drawback in the proposed approach we extract a set ok key features from the tweets that are incrementally grouped through a clustering algorithm relying on co-occurences of bursty textual features (e.g., hashtags and words). To find the sentiments of the detected topics, a classifier that considers the overall sentiment of all the tweets in the topics is proposed.

The proposed approach overcomes the limitations of current state-of-the-art that mainly focus on extracting the sentiment from each individual tweets. In fact, even if sentiment classification is a widely studied research line, sentiment analysis of topics is not investigated adequately. Current research focused mainly on performing sentiment analysis at document and sentence level. Document-level sentiment analysis extracts the overall polarity of a whole document while sentence level focuses on fine-grained analysis where each sentence is treated as an independent element and the approach is based on the assumption that the sentence refers to one opinion. Accordingly, traditional sentiment analysis focuses on classifying the overall sentiment expressed in a text (that could be a document or a sentence) without specifying what the sentiment is about. This may not be enough if the text is simultaneously referring to different topics, possibly expressing different sentiments towards different topics.

The proposed solution is designed as a novel hybrid sentiment classification model combining sentence and topic-based approaches. The main rationale behind this choice is that predicting an overall score for a tweet is not suitable since a tweet can mention different topics. Therefore, it is more effective to perform sentence-level sentiment of each tweet and multi-topic sentiment classification, predicting different ratings for each topic discussed in the tweet rather than an overall rating.

In the following are reported some examples to make clearer many of the above concepts.

The sentiment of words used in a tweet are often dependent on the topic/s of that tweet. For example let's consider the tweet "*Happy!Remdisivir is working symptoms disappeared fortunately as I dont trust in vaccines*". The tweet belongs to two topics, the "*Medicine*" topic and the "*Vaccine*" topic. The tweet expresses a positive sentiment towards the "Medicine" topic while it expresses a negative sentiment toward the "Vaccine" topic. If we would consider the general sentiment of the tweet would be wrongly fully positive.

Another example of tweet with controversial sentiment is the following: "*Stay safe from Coronavirus. There is currently no antiviral treatment or vaccine to prevent COVID-19*". In fact, the bigrams "*stay safe*" could determine a positive sentiment whereas the overall sentiment of the tweet is determined by the absence of a treatment or vaccine to prevent COVID-19.

To address the above challenges and issues, the paper introduces the *Topic-specific Sentiment Classifier* (TSC) algorithm, which is only trained on tweets of the same topic.

3.2 The Algorithm

The *Topic-specific Sentiment Classifier* (TSC) algorithm is formulated as a pipeline model combining two main machine learning methods: an unsupervised learning approach for topic modeling and a supervised classification algorithm to perform sentiment analysis with reference to the topics.

For detection of topics from streaming data, an online clustering algorithm is proposed. The algorithm, given a timestamp t, first extracts a set of features from the tweets for the k preceding temporal values of t and model each feature as a time series. Then, an ad-hoc peak detection method analyzes the time series to find peaks/bursty. After that, a clustering algorithm will group the bursty textual features to identify trending topics of discussion related to the COVID-19 outbreak. The clustering approach is based on co-occurences of the keywords in the tweets. A preliminary version of the topic detection approach has been presented in [4].

After the topics have been detected, the classifier analyses the tweets in the topics and the set of sentiment features. Sentiment classification is formulated as a multi-class classification problem, where the textual features are classified into one of three or more classes. In this work, we focus on the single-label version of the multiclass classification problem in which instances are associated with exactly one element of the label set. In the proposed approach we apply a strategy for reducing the problem of multiclass classification to multiple binary classification problems relying on the well-known One-vs-All (OVA) strategy.

4 Experimental Evaluation

4.1 Twitter Datasets

The approach has been validated on a set of tweets posted in the United States. Specifically, two different datasets has been used. The first dataset is the one collected within the CoronaVis project [7] and accessible from the Github repository (https://github.com/mykabir/COVID19). The dataset consists of over 200 million tweets related to COVID-19 posted by 30.070 unique users in the period March 2020 - June 2021. Among the keywords used to filter COVID-19 tweets there were *corona, pandemic, lockdown, quarantine, virus, pneumonia, outbreak,* etc.

Since some days in the faced period were missing due to connectivity issues, the dataset in [7] has been integrated with tweets collected in the GeoCOV19Tweets dataset [8]. GeoCOV19Tweets consists of 675,104,398 tweets (available from the web site https://ieee-dataport.org/open-access/coronavirus-covid-19-geo-tagged-tweets-dataset) and contains IDs and sentiment scores of the geo-tagged tweets related to the COVID-19 pandemic.

While the data provided in the CoronaVis dataset [7] is already preprocessed, the GeoCOV19Tweets set required a preliminary preprocessing step to remove retweets, punctuation, stop words and emojis. Moreover, stemming has also been performed.

4.2 Sentiment About COVID-19 in US

Sentiment Scores. To compute the sentiment of the topics and, thus, determine the score of the sentiments, we refer to a set of widely adopted metrics used for sentiment analysis. In particular, we used the package in [5], which classifies each sentiment as either positive (+1 score), negative (−1 score) or neutral (0 score).

Let N be the total number of tweets containing a keyword k, N_{pos}, N_{neg}, and $N_{neutral}$ the number of positive, negative, and neutral tweets regarding k, respectively.

Polarity is the ratio between the number of positive tweets and the number of tweets that express a sentiment about k.

$$polarity = \frac{N_{pos}}{N_{pos} + N_{neg}} \tag{1}$$

Subjectivity gives the fraction of not neutral tweets with respect to the total number of tweets.

$$subjectivity = \frac{N_{pos} + N_{neg}}{N} \tag{2}$$

Positive is the ratio between the number of positive tweets and the total number of tweets.

$$positive = \frac{N_{pos}}{N} \tag{3}$$

Negative is the ratio between the number of negative tweets and the total number of tweets.

$$negative = \frac{N_{neg}}{N} \tag{4}$$

Results. In the rest of the section we present the results obtained by using the topic-specific classifier, aiming at analyzing people sentiment concerning COVID-19 as expressed in Twitter.

We start the analysis by showing in Table 1 the top 12 topics of discussion detected from the considered dataset of tweets. Table shows for each topic a brief description, the top frequent content features and top frequent sentiment features.

As can be noted from Table 1, the top topics of discussion span from the initial days of the outbreaks with conversations about the novel virus appeared in China and spreading worldwide, characterized of sentiments of fear, alert and stress caused by COVID-19, due to its quick spread, and continue with a variety of topics of different natures and expressing a variety of sentiments. We

Table 1. Top 12 topics of discussion about COVID-19 in US.

Topics	Content features	Sentiment features
Covid-19 outbreak	2019-nCov, novel coronavirus, corona, Covid-19, outbreak, China, pandemic	spread, emergency, fear, worry, alert, stress, scared
Lockdown	measure, lockdown, quarentine, prevention, stop	ability, assurance, alarm, abuse, burden
Cases	case, cases, confirmed suspect, lag	advantage, accept, accuse accusation, crisis, bailout, fear
Death	dead, died, kill, death, sick, ill	fear, alarm, worry, relieved, stress
Hospital overwhelmed	hospital full, nursing, doctors, ill, ventilators, no bed, shed	help, care, alarm, alert, worry, effective, manage, readiness
Quarantine	home, close, lockdown, alone, inside, stay inside	benefit, assure, respect, follow, escape, avoid, anti, stressed
Medical supplies	mask, disinfection, protection, shortage, gloves, gel	effective, mandatory, annoying, tired, difficult
Virus research	detection, research, laboratory, treatment, vaccine	hope, efficacy, confident, negative, difficult
Testing center	result, test, tested, positive, negative, quantitative, symptoms	pain, relief, worried, scared
Vaccines	vaccine, immune, antibody, mouse, vaccination, animal, protective, shot	against, pro, no-vax, skeptical, effective, anti, danger
Remote Work	remote, work, remote working, office close, cancel school, home	admit, agree, mandatory, annoying, tired, difficult
Crisis	emergency, crisis, pandemic, policy, strategy, economic, loss	prevent, capacity, effectively, manage, strengthen, support, readiness, anger
Medicines	Remdesivir, Tocilizumab, Clorochine, treats, hydroxychloroquine	efficacy, concern skeptical, danger

can observe that there are topics discussing the preventive measures adopted to fight COVID-19 and limit its diffusion like *Lockdown, Quarantine, Medical supplies* characterized by sentiments of approval/ disapproval, efficacy or inadequacy of the adopted measures; the curative measures used to fight COVID-19 (e.g., hospital overwhelmed, testing center, drugs) expressing again agreement/disagreement, efficacy but also fear, worry, skepticism; similar opposite sentiments characterize the topics arisen around the research efforts started to study and fight the pandemics (e.g., vaccines, virus research); topics concerning the virus tracking and surveillance like *Cases, Deaths, Testing Center* with sentiments of fear, worry, alarm, disappointment; other topics discussing about the consequences brought in people life like *Remote working, Crisis* with negative sentiments of worry and fear. The topic *Medical supplies* concerns tweets about the importance of facial masks and gloves as prevention measures to reduce the outbreak and also their shortage in several countries. Tweets about quarantining people infected or suspected to have COVID-19 are grouped in the topic *Quarantine*.

In the following of the section we present a sentiment analysis study of some of the top interesting topics of discussion shown in Table 1. In particular, we show the sentiment of people with respect to two main categories of topics detected from the tweets:

1. the preventive measures adopted to limit the diffusion of the pandemic (e.g., vaccines, masks, lockdown);
2. the curative measures used to fight COVID-19 (e,g., hospital overwhelmed, medicines, testing center, quarantine).

Figure 1 shows the subjectivity of the curative measures as discussed in the three topics *Medicine, Quarantine and Hospital Overwhelmed.* We first show the subjectivity index to understand how many users express a sentiment. Figure 1 shows that for the topic *Hospital Overwhelmed* the subjectivity values is rather high during the first weeks of observation, meaning that people devoted a lot of attention on the topics with a high percentage of people expressing a sentiment about the curative measures adopted that is not neutral. Then, subjectivity gradually attenuates meaning it lose attention after the first active phase.

The topic *Medicine* exhibits very low subjectivity at the beginning, but after the initial phase of the epidemic, the topic gained attention reaching a peak early days of April and remaining rather high till June. After that, it loses the interest of people. The subjectivity of the topic *Quarantine* during the first days of the epidemic outbreaks is rather low, while it grows over the weeks reaching a peak around mid May. After that, users loose interest in the topic *Quarantine* whose subjectivity decreases till reaching a very low value during the last weeks of the observation.

Figure 2 illustrates how the *polarity* index varies for the three topics over the weeks. Recall that polarity is the ratio between the number of positive tweets about a given topic and the number of tweets that express a sentiment about that topic. In general, the *polarity* values are rather low for all the topics over the weeks. This means that the number of tweets expressing a positive sentiment is inferior than that expressing a negative one. The negative sentiment towards those topics could be explained with the fact that people is not happy with the measures used to fight the epidemics. The polarity of *Hospital Overwhelmed* is very negative throughout the period. The polarity of *Medicine* is always rather low except for a peak around May. The polarity of *Quarantine* is always low even if presents a growing trend and it has a very high peak around October when the epidemic resumed strong vigor. Low polarity values are confirmed by the *positive* and *negative* indexes, showed in Figs. 3 and 4. For example, from Fig. 4, we can observe that for *Hospital Overwhelmed* at the beginning of the observation the sentiment is very negative and the negative sentiment prevails till end May, then it decreases and starts to be again very negative reaching end of October. For what concerns the topic *Quarantine* the negative sentiment grows over the weeks till reaching a peak around mid May, then it gradually decreases till reaching low negative values starting from September when the virus started to circulate again and, as thus, we can see from Fig. 3 that the positive index has an important peak in October.

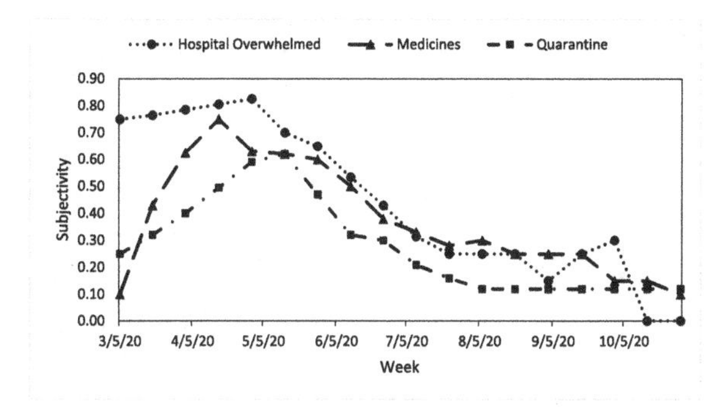

Fig. 1. Weekly Subjectivity index for the curative measures.

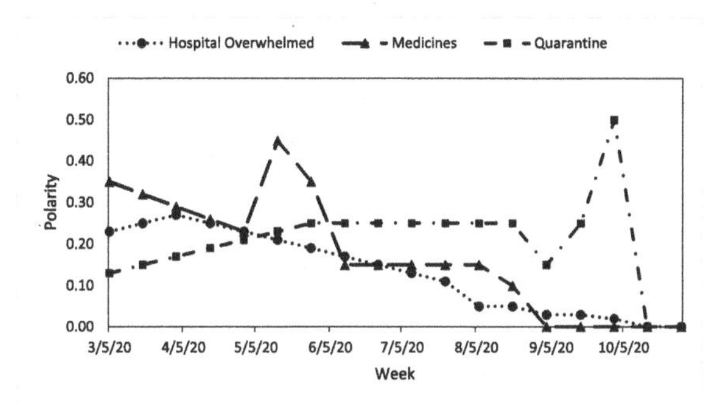

Fig. 2. Weekly Polarity index for the curative measures.

For the preventive measures we considered the following topics: *Lockdown, Vaccine, Medical supply.* Figure 5 shows that for the preventive measures used to fight COVID-19 the subjectivity values follow a fluctuating trend over the weeks, meaning that people attention towards those topics changed during the target period. In particular, Fig. 5 points out that a high percentage of people expresses a sentiment about vaccine that is not neutral. In fact, the subjectivity for such a topic is rather high especially during the last weeks of the observed period.

Figure 6 illustrates how the *polarity* index varies for each preventive measures over the weeks. Figure 6 shows that, in general, the *polarity* values are rather low for the topics *Lockdown* and *Medical supply* over the weeks, except in the period spanning from mid April till end of July. This means that the number of tweets expressing a positive sentiment is inferior than that expressing a negative one. Differently, the topic *Vaccines* has an opposite trend. Low polarity values are confirmed by the *positive* and *negative* indexes, showed in Figs. 7 and 8. For example, from Fig. 8, we can observe that for *Lockdown* at the beginning of the observation the sentiment is very negative, about 0.6, then it decreases and becomes rather positive, after that, it starts to be again very

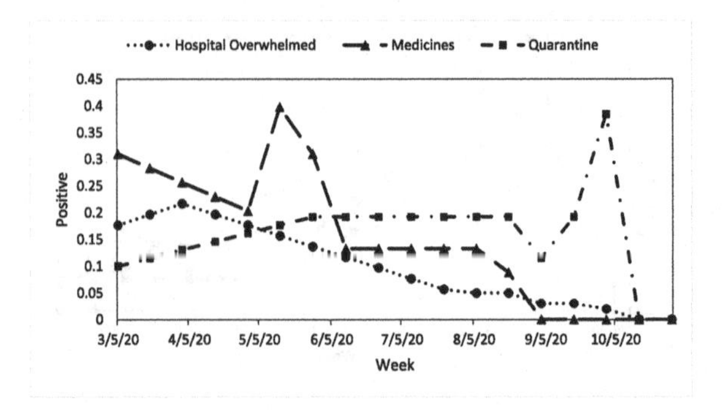

Fig. 3. Weekly Positive index for the curative measures.

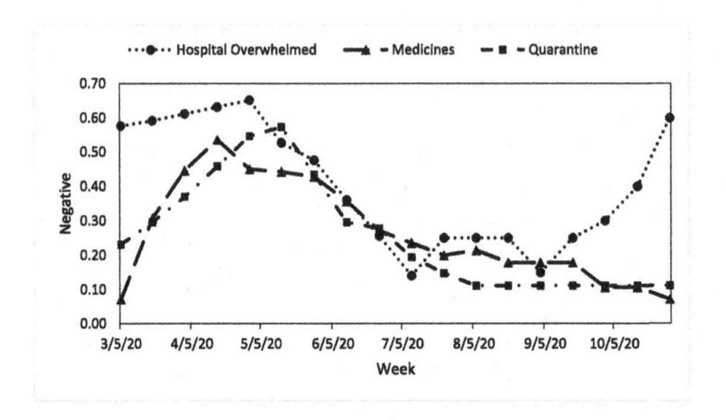

Fig. 4. Weekly Negative index for the curative measures.

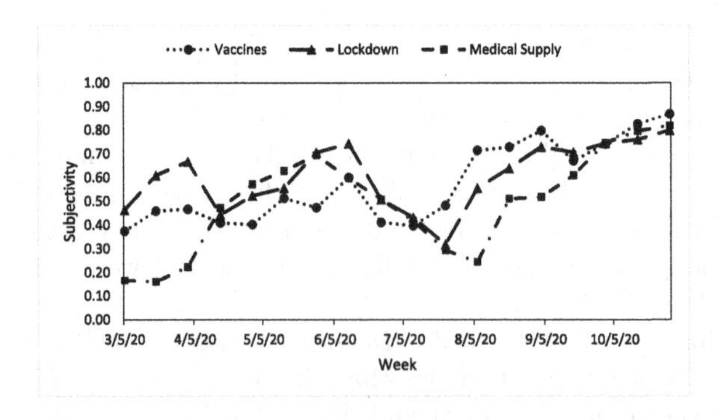

Fig. 5. Weekly Subjectivity index for the preventive measures.

negative around August. The sentiment towards *Lockdown* becomes again quite positive during the last week of observation. Conversely, the topic *Vaccines* first exhibits a negative sentiment but around July the sentiment started to change. In particular, during the last weeks of the observation, when the first promis-

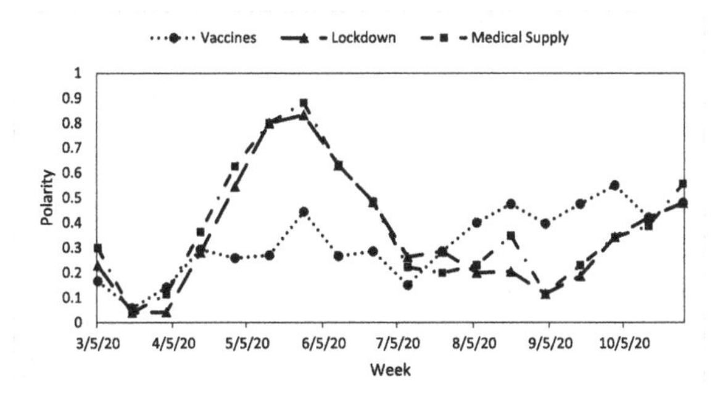

Fig. 6. Weekly Polarity index for the preventive measures.

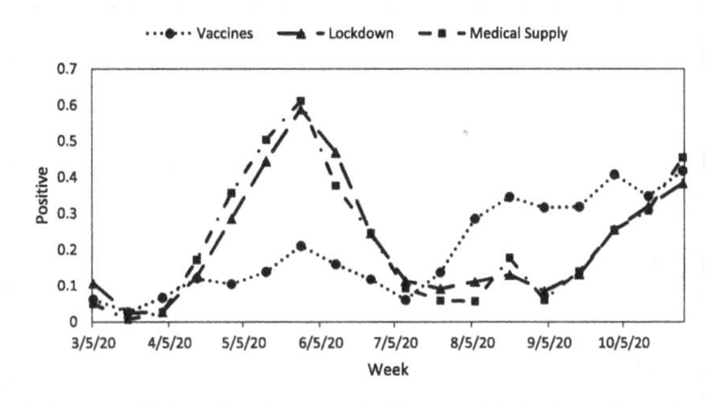

Fig. 7. Weekly Positive index for the preventive measures.

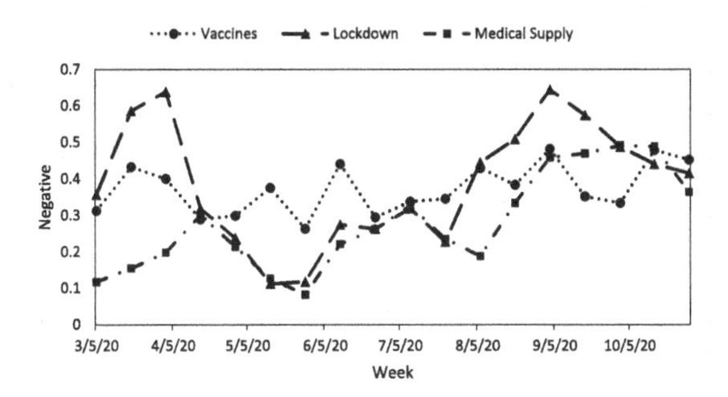

Fig. 8. Weekly Negative index for the preventive measures.

ing outcomes about vaccines testing have been released, the people started to change idea towards vaccines and, consequently, the prevalent sentiment of the discussion arisen around vaccines became positive.

The topic *Medical supplies* exhibits a very low subjectivity at the beginning, than slowly increases and after end of May it decreases again, reaching zero during the summer months when the epidemics was under control. After that, it sharply increases in correspondence of end of August. This means that at he beginning of the observation few people expressed a sentiment about the medical aid especially when the wave of the epidemics during the summer was under control. At a certain point, when the epidemics resumed vigorously, the discussions concerning the masks increased and, as can be seen from the value of the Negative index, the polarity is very negative. This could indicate a certain distrust on the part of people in the use of masks as main tool to fight COVID-19.

5 Conclusions

In this work a topic-level sentiment analysis approach based on topic modeling and supervised machine learning classification models has been proposed. The proposed approach supports scalable and dynamic topic detection over streaming social media data and performs sentiment analysis at topic-level.

To discover relevant COVID-19 topics of discussion from Twitter, the work presented an approach that combines identification of bursty features and online clustering. Topics are incrementally obtained by grouping the bursty textual features based on the co-occurrence in the tweets. Each cluster obtained is then associated to a topic. After a topic has been detected, a multi-label classifier computes its sentiment. The novelty of the proposed approach is that it works at the topic level to extract sentiments from social media data, differently to literature approaches that instead extract the sentiments at the tweet-level.

The proposed TSC approach has the ability to detect multiple topics and associated sentiments from streaming social media posts. Results, performed over a real-world dataset of tweets, shown the feasibility of the method that is able to detect a large number of relevant COVID-19 topics and extract people sentiment toward such topics. Results also shown that the TSC algorithm outperformed state-of-the-art baseline models in different scenarios and by using different classifiers.

Acknowledgments. This work has been partially supported by project SERICS (PE00000014) under the NRRP MUR program funded by the EU - NGEU.

References

1. Abd-Alrazaq, A., Alhuwail, D., Househ, M., Hamdi, M., Shah, Z.: Top concerns of tweeters during the COVID-19 pandemic: infoveillance study. J. Med. Internet Res. **22**(4), e19016 (2020)
2. Blei, D.M., Ng, A.Y., Jordan, M.I., Lafferty, J.: Latent Dirichlet allocation. J. Mach. Learn. Res. **3**, 2003 (2003)
3. Comito, C.: How covid-19 information spread in us the role of twitter as early indicator of epidemics. IEEE Trans. Serv. Comput. 1–1 (2021). https://doi.org/10.1109/TSC.2021.3091281

4. Comito, C., Falcone, D., Talia, D.: A peak detection method to uncover events from social media. In: 2017 IEEE International Conference on Data Science and Advanced Analytics, DSAA, pp. 459–467. IEEE (2017)
5. Feinerer, I., Hornik, K., Meyer, D.: Text mining infrastructure in R. J. Stat. Softw. **25**(5), 1–54 (2008)
6. Ferrara, E.: What types of covid-19 conspiracies are populated by twitter bots? First Monday (2020). https://doi.org/10.5210/fm.v25i6.10633
7. Kabir, M.Y., Madria, S.: Coronavis: a real-time covid-19 tweets analyzer (2020)
8. Lamsal, R.: Coronavirus (covid-19) geo-tagged tweets dataset
9. Li, L., et al.: Characterizing the propagation of situational information in social media during covid-19 epidemic: a case study on WEIBO. IEEE Trans. Comput. Soc. Syst. **7**(2) (2020). https://doi.org/10.1109/TCSS.2020.2980007
10. Ordun, C., Purushotham, S., Raff, E.: Exploratory analysis of covid-19 tweets using topic modeling, umap, and digraphs (2020)
11. Kouzy, R., et al.: Coronavirus goes viral: Quantifying the covid-19 misinformation epidemic on twitter. Cureus **12**(3), e7255 (2020)
12. Rashid, M.T., Wang, D.: Covidsens: A vision on reliable social sensing for covid-19 (2020)
13. Schild, L., Ling, C., Blackburn, J., Stringhini, G., Zhang, Y., Zannettou, S.: "go eat a bat, chang!": an early look on the emergence of sinophobic behavior on web communities in the face of covid-19 (2020)
14. Shahsavari, S., Holur, P., Tangherlini, T.R., Roychowdhury, V.: Conspiracy in the time of corona: automatic detection of covid-19 conspiracy theories in social media and the news (2020)
15. Sharma, K., Seo, S., Meng, C., Rambhatla, S., Liu, Y.: Covid-19 on social media: analyzing misinformation in twitter conversations (2020)
16. Singh, L., et al.: A first look at covid-19 information and misinformation sharing on twitter (2020)

Lightweight Weight Update for Convolutional Neural Networks

Feipeng Wang[1,2], Kerong Ben[1], Xian Zhang[1](✉), and Meini Yang[1]

[1] College of Electronic Engineering, Naval University of Engineering, Wuhan, China
1920191172@nue.edu.cn

[2] School of Computer and Big Data Science, Jiujiang University, Jiujiang, China

Abstract. Convolutional neural networks are usually composed of convolutional layers and pooling layers. Pooling operations effectively control the weight update of convolutional neural networks. The existing pooling operations result in a large number of weight parameter updates of convolutional neural networks, causing large memory usage. In this paper, a pooling operation called ApproxM is proposed to address this problem. The proposed pooling operation is a simple and similar to median pooling. It takes the mean of multiple values near the median as the pooling result, and CNNs only update the weights of these values during the back propagation. Finally, extensive experiments on benchmark datasets demonstrate that the proposed pooling operation achieves top-1 results of 92.65% and 68.24% and top-5 results of 99.84% and 91.31% model test accuracy on Cifar-10 and Cifar-100 based on ResNet-20, respectively, and the corresponding number of weight updates in an 8×8 pool is 4, which is better than other pooling techniques in the experiments.

Keywords: Approximate Median Pooling · Weight Update · Image Recognition · Convolutional Neural Networks

1 Introduction

Despite the unprecedented success of deep neural networks in various applications including computer vision [1–3], natural language processing [4] and robotics [5], understanding the role of various operations in network architecture is still a challenging problem. While the use of some operations exhibit impressive performance in different domains, the reasons for the performance improvement are often not well understood.

Researchers [6] have been investigating and explaining power and efficient operations applicable to deep neural networks, which include convolutions, normalizations, and activation functions. However, they have not paid much attention to pooling operations despite their simplicity and effectiveness in aggregating local features. The number of weight updates is critical; too much or too little weight updates, its cost performance difference is larger. The number of weight updates is determined by several factors in deep neural networks. To design an

S. Zhang et al. (Eds.): BigData 2023, LNCS 14203, pp. 108–118, 2023.
https://doi.org/10.1007/978-3-031-44725-9_8

efficient operation to reasonably update weights, variants of convolution operations or activation functions with different functions are widely adopted, and they have been explained accordingly. However, few researchers have explained the role that pooling manipulation plays in weight updates.

To alleviate the suboptimality of the pooling operations and rationally explain them, this paper proposes a novel pooling operation termed approximate median pooling (ApproxM Pooling), which is a module with an adjustable number of weight updates, replacing Max Pooling and Average Pooling operation typically used. The proposed module favors finding the optimal weight update during back propagation. This allows us to better understand the role of the Pooling operations. The contributions of this paper are summarized as follows:

- This paper explains the limitations of the commonly used Max Pooling and Average Pooling operation, and points out the importance of updating numbers during weight updating, which is still under-explored in designing neural network architectures.
- The proposed ApproxM Pooling, a module with an adjustable number of weight updates, that can find the optimal weights updates. ApproxM Pooling calculates the average of the values around the median of the feature values to obtain the pooling results and also determining the location of the back propagation update weights to optimization across the entire network.
- Extensive experiments demonstrate that the model with ApproxM Pooling outperforms the baseline algorithms on multiple datasets and network architectures in the image classification tasks. It also exhibits desirable trade-offs between accuracy and weight update cost.

This paper is organized as follows. Section 2 presents existing related works, and Sect. 3 introduces the motivation for optimizing the pooling operation. The experiments details and results of ApproxM Pooling are described in Sect. 4. Last, Sect. 5 concludes this work and discusses the future works.

2 Background and Related Works

Neural Network Architecture. A typical deep convolutional neural network (CNN) architecture with image classification usually has two parts: feature extraction and classification recognition. The feature extraction part usually contains convolution layer, activation function, and pooling operation layer. The commonly used convolution kernel has 1×1, 3×3, 5×5, the activation function has ReLu, and the pooling operation has Max Pooling and Average Pooling. The classification recognition part is generally a fully connected layer and a classification function, such as the Softmax function. Classical architectures include AlexNet [7], VGG [8], ResNet [3], Wide Residual Networks (WRN) [9], etc.

Weight Update Strategy. The convolutional layer shares the weight parameters through the convolution kernel, which greatly reduces the number of weight parameters compared to the fully connected layer. The activation function plays a certain shielding against some weaker neuronal inputs. Dropout randomly deactivates some neurons. The pooling operation also disabled some neurons while downsampling.

Pooling and Its Weight Update Characteristics. LeCun et al. proposed and used Average Pooling [10,11], who used Max Pooling [12] in their 2007 article entitled Sparse Feature Learning for Deep Belief Networks. For more than 30 years, a large number of researchers have put forward the pooling operation that can effectively solve the problems they have encountered in their research work. In popular operations, (1) update single weight parameter, such as Max Pooling [12], Global Max Pooling [13], Row-wise Max Pooling [14], Stochastic Pooling [15]; (2) update partial weight parameters, such as Mixed Pooling [16]; (3) update all weight parameters, e. g, Average Pooling [11], Rank-based Average Pooling [17]. The most commonly used pooling operations include Max Pooling [12] and Average Pooling [11], one is to update one weight parameter, the other is to update all the weight parameters. The ApproxM Pooling operation proposed in this paper belongs to updating partial weight parameters. Figure 1 shows the functions of Max Pooling [12], Average Pooling [11] and ApproxM Pooling operations during forward propagation and back propagation.

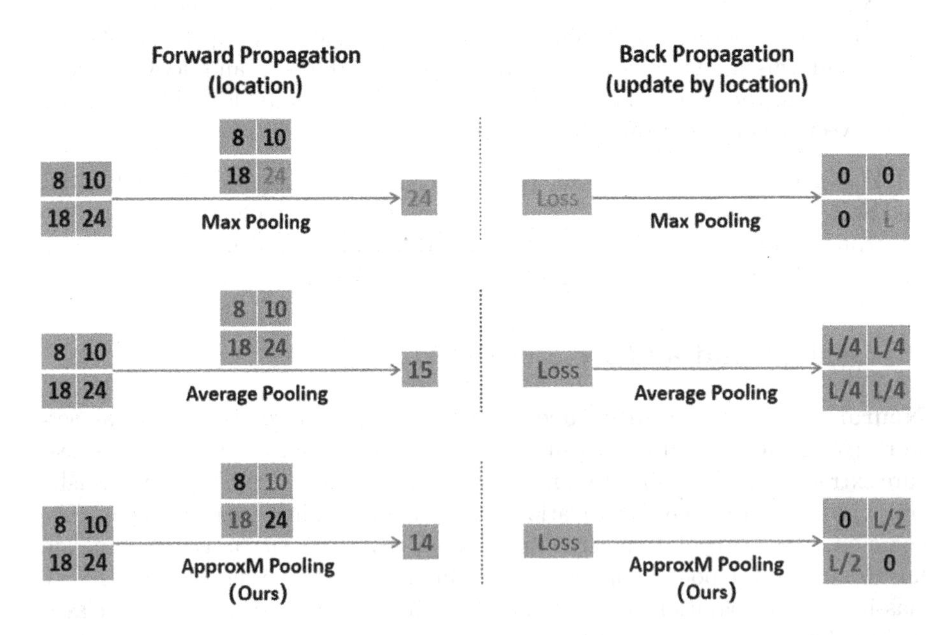

Fig. 1. The function of the pooling operations in forward propagation and back propagation.

3 Proposed Method

Motivation. "A single red flower in the midst of thick foliage" can be quickly captured by the human eye, because the color difference between red and green blocks is large, resulting in a transition boundary, and the human eye quickly captures this boundary. The feature points in the image are mainly distributed at the boundaries between color patches. Feature extraction by human vision is fast and robust. A pixel is perceived more easily because the sum of the contrast of its color with the color of the surrounding pixels is minimal and the pixel itself is relatively bright, which is generally in the boundary transition region.

The pooling operation in CNNs is very similar to this process used by human vision to extract features. Pooling operations are commonly believed to have the following effects: (1) downsamping; (2) dimensionality reduction to reduce the number of parameters; (3) realize nonlinearity; (4) expand the receptive fields; (5) Realize translation invariance, rotation invariance and scale invariance. However, these understandings are from the perspective of forward propagation. Their main role can mostly be replaced by convolutional layers. If from the perspective of back propagation, that is, when the parameter is updated by gradient, the pooling layer plays the role of the activation function in the forward propagation. Whether the corresponding weight parameter is updated or not will play a switch role, and then the update amount will be calculated according to the current loss.

Minimum Contrast Sum. To calculate the minimum value of the sum of the contrast between the pixels and the surrounding pixels is actually approximately finding the median value of a set of values. For example, a set of values in Fig. 2 8, 10, 18, 24, then the sum of the contrast of 8 is 28($|8\text{-}10| + |8\text{-}18| + |8\text{-}24|$), the sum of the contrast of 10 is 24($|10\text{-}8| + |10\text{-}18| + |10\text{-}24|$), the sum of the contrast of 18 is 24($|18\text{-}8| + |18\text{-}10| + |18\text{-}24|$), and the sum of the contrast of 24 is 36($|24\text{-}8| + |24\text{-}10| + |24\text{-}18|$). The last minimum value of the sum of the contrast is 24, and the corresponding pooling value is 18. If this set of values is sorted, its 10,18 is exactly the closest value to their median value. In the above example, the approximation process between the minimum contrast sum and the median is shown in Fig. 2.

Approximate Median Pooling Operation. The median is the value that takes the middle position, or it is the mean of two values at the middle position. The pool size is generally even, and the value corresponding to minimum value of the sum of the contrast in the pool is a number close to the median value. We propose a novel approximate median pooling operation termed ApproxM Pooling. The ApproxM Pooling takes the average of multiple numbers in the pool near the median as the pooling result.

In the forward propagation, the ApproxM Pooling takes the average of the ocount features on either side of the median as the pooling result, locates the

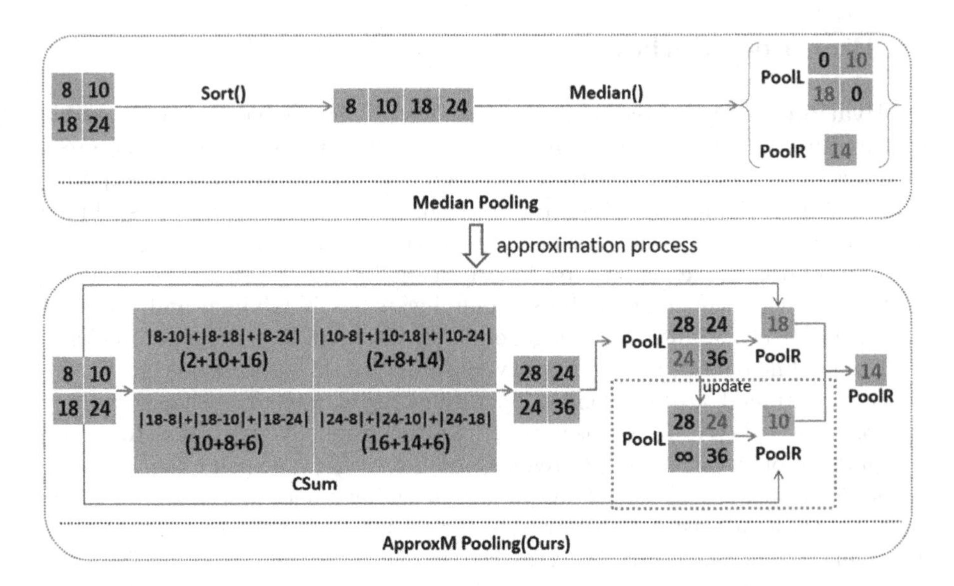

Fig. 2. The function of the pooling operations in forward propagation and back propagation.

ocount features to prepare for updating weight parameters during the back propagation, and plays the role of switching on whether the weight parameter is update or not. The procedure of the approximate median is shown in Fig. 2. The core algorithm steps are as follows:

Step 1: Calculate the contrast sum;
Step 2: Locate the minimum contrast sum;
Step 3: Obtain the pooling positioning and results according to the minimum contrast sum positioning;
Step 4: Determine whether the pooling numbers has been reached. If the pooling numbers has been reached, jump to Step 6; Otherwise, go to the next step;
Step 5: The current minimum contrast sum is set to positive infinity, jump to Step 2;
Step 6: Output pooling positioning and results.

The lower part of Fig. 2, with inputs of 8, 10, 18, 24, shows the average result of taking two values as an example of pooling results. According to the above algorithm, the corresponding contrast sum calculated in step 1 is 28, 24, 24, 36. Step 2 locate the minimum contrast sum to 24. Step 3 based on the result of Step 2, the corresponding result is 18. Step 4 determine the number of pooling values. In this case, the number of pooling values is 2. Step 5 set the value of the current minimum contrast sum to infinity (change the value of only one position at a time). Then jump to step 2 to locate and obtain 10, and finally output the position of 10, 18 and their average result 14.

4 Experiments

To evaluate the performance of the ApproxM Pooling, we apply it to a variety of image recognition datasets: Cifar-10 and Cifar-100 [18]. The CNN architectures used for comparison are the ResNet Residual Networks [3].

Experiments Setup. In all of our experiments, we compare the CNN models trained with or without the ApproxM Pooling. For the same deep CNN architecture, all the models are trained from the same weight initialization. If not specified, all models perform data augmentation on the training data prior to training. Firstly, each side of the image is padded by 4 pixels. And then the images of Cifar-10 and Cifar-100 datasets are randomly cropped with 32×32, 32×32, respectively. Finally, the images are flipped horizontally. We evaluate top-1, top-5 accuracy rates in the format "mean $\pm$ std" based on 10 runs on ResNet Residual Networks [3].

Cifar-10 and Cifar-100 contain 60,000 color images of size 32×32 pixels, respectively. Cifar-10 is divided into 10 classes with 6,000 images per class. This contains 50,000 for training, forming five training batches of 10,000 images each; another 10,000 for testing, forming a separate batch. The data of the test batch are taken from each of the 10 classes, 1000 images are taken randomly in each class. The remaining ones are randomly arranged to form the training batch. Therefore, all the types of images in a training batch are not necessarily the same number, but overall the training batch, it has 5,000 images in each class. The Cifar-100 is very similar to the Cifar-10, but it has 100 classes, each containing 600 images. Each class has 500 training images and 100 test images. 100 classes in the Cifar-100 are divided into 20 superclasses. Each image has a "fine" label (the class that it belongs to) and a "rough" label (the superclass that it belongs to).

The 8, 14, 20-layer networks for ResNet are used in the same way as ResNet. The training procedure follows ResNet. SGD is used as an optimizer with a learning rate of 0.1, a momentum of 0.9, and a weight decay of 0.0005. Especially, the learning rate of ResNet start from 0.1. The learning rate of the ResNet is divided by 10 after the 100th, 175th, and 250th epochs. The ResNet is stopped training by the 300th epoch.

Main Results. ApproxM Pooling mainly has five different weight update numbers in Fig. 3. We compare their effects on test accuracy with those of Max Pooling and Average Pooling. We train ResNet-20 on the Cifar-10 and Cifar-100 datasets and perform different pooling operations and weight update numbers. In Fig. 3, ApproxM Pooling with update weights of 2 and 4 has the same test accuracy performance as Average Pooling, while Average Pooling has 64 weight updates. Although the Max Pooling weight update is only 1, its test accuracy is lower than that of ApproxM Pooling with weight update 2, especially on Cifar-10.

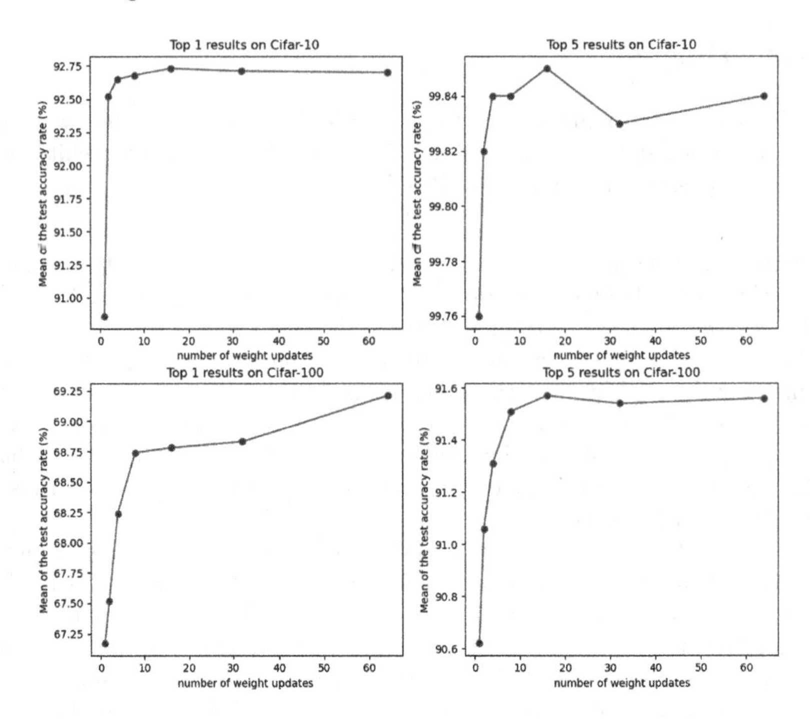

Fig. 3. Test accuracy (%) of different pooling operations and weight update numbers on Cifar based on ResNet-20. The update weights number of 1 and 64 are Max Pooling and Average Pooling respectively, and the rest (2/4/8/16/32) are ApproxM Pooling.

We consider the effect of increasing the number of weight update parameters on the test accuracy. The test accuracy results of ApproxM Pooling with different weight update parameters, Max Pooling, Global Max Pooling and Average Pooling on Cifar based on ResNet-20 are shown in Table 1. In Table 1, ApproxM Pooling (2) adds a weight update parameter relative to the Max Pooling, and ApproxM Pooling (4) adds two weight update parameters relative to ApproxM Pooling (2). Each additional weight update parameter increases the test accuracy by 0.35 and 0.36 respectively. Taking the Max Pooling test accuracy as the starting point, the accuracy change caused by increasing the number of weight updates is calculated, and the results are shown in Fig. 4.

Table 1. The Top 1 test accuracy results of different pooling operation on Cifar based on ResNet-20. The n in ApproxM Pooling(n) is the number of weight update parameters.

Pooling Operation	Cifar-10	Cifar-100
Max Pooling	90.86 ± 0.78	67.17 ± 0.60
Global Max Pooling	91.70 ± 0.42	67.07 ± 0.47
ApproxM Pooling(2)	92.52 ± 0.21	67.52 ± 0.27
ApproxM Pooling(4)	92.65 ± 0.28	68.24 ± 0.32
ApproxM Pooling(8)	92.68 ± 0.14	68.74 ± 0.30
ApproxM Pooling(16)	92.73 ± 0.21	68.78 ± 0.23
ApproxM Pooling(32)	92.71 ± 0.17	68.83 ± 0.26
Average Pooling	92.70 ± 0.14	69.21 ± 0.48

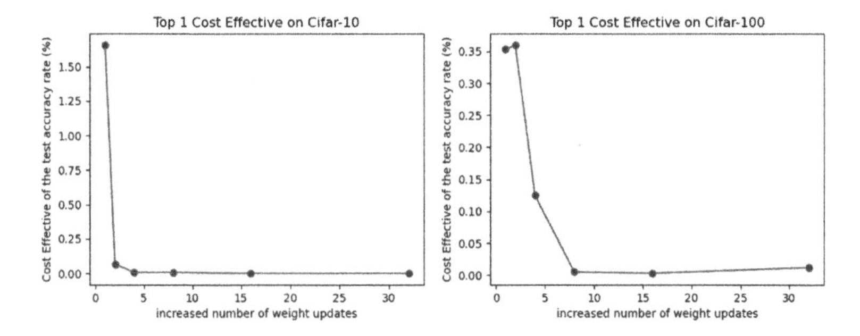

Fig. 4. The effect of weight update parameter number on test accuracy.

Comparison with Max and Average Pooling. We compared the effect of ApproxM Pooling with Max Pooling and Average Pooling on test accuracy. We train ResNet-20 on the Cifar-10 and Cifar-100 datasets and perform different pooling operations. In Fig. 5, ApproxM Pooling with update weights of 4 has the same test accuracy performance as Average Pooling, and is obviously better than the Max Pooling.

We compared the effect of ApproxM Pooling with Max Pooling and Average Pooling on test accuracy. We train ResNet-8, ResNet-14 and ResNet-20 on the Cifar-100 dataset and perform different pooling operations. In Fig. 6, ApproxM Pooling, Max Pooling and Average Pooling show the same performance law on ResNet-8, ResNet-14 and ResNet-20, indicating that ApproxM Pooling has a stable effect on different network structures.

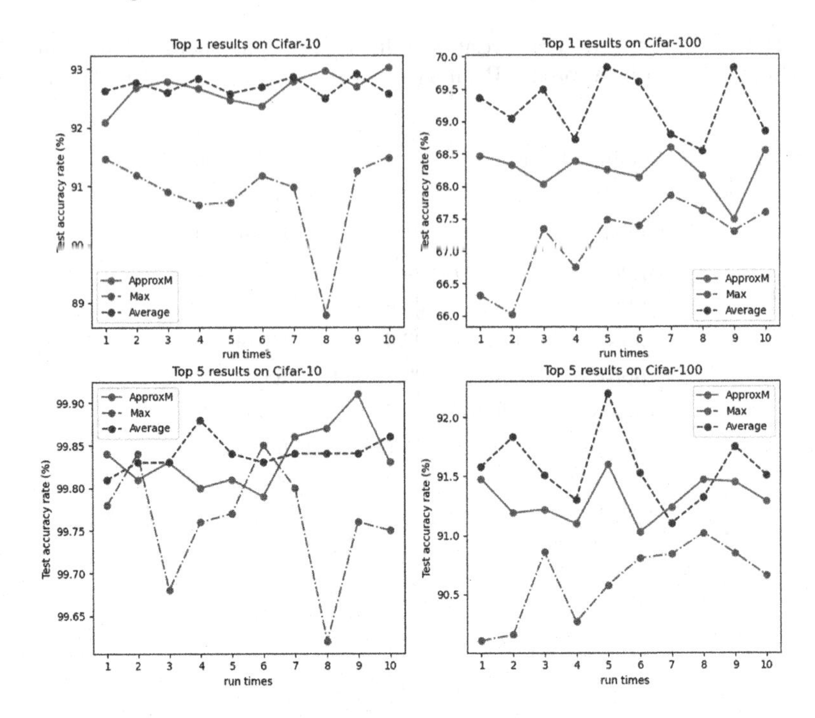

Fig. 5. Test accuracy (%) of different pooling operations and weight update numbers on Cifar based on ResNet-20. ApproxM Pooling uses the update weight of 4.

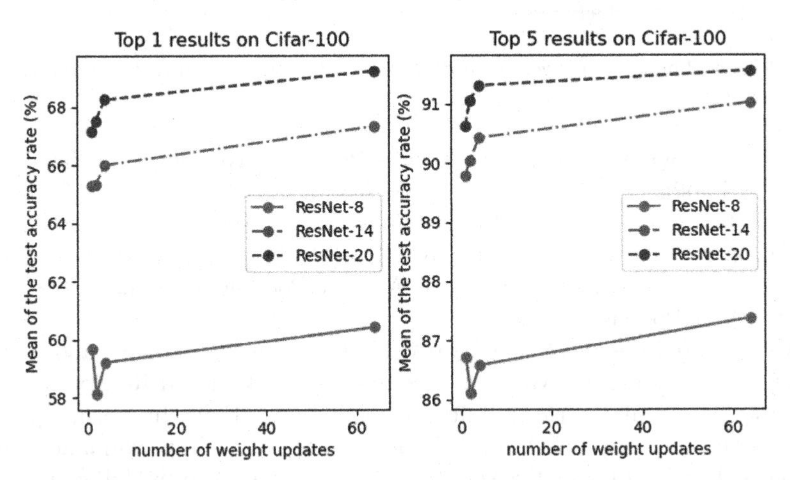

Fig. 6. Test accuracy (%) of different pooling operations on Cifar-100 based on ResNet-8, ResNet-14 and ResNet-20. ApproxM Pooling uses the update weight of 4.

5 Conclusions

In this paper, a new method called ApproxM Pooling is proposed. ApproxM Pooling only needs to simply take the mean of multiple values near the median as the pooling result, and only update the weights of these values during back-propagation. Extensive experiments on image classification for different network architectures on Cifar-10 and Cifar-100 validate the excellence of the proposed pooling method. ApproxM Pooling obtains 92.65% and 68.24% top-1 result on Cifar-10, Cifar-100 based on ResNet-20, respectively, but the corresponding number of weight updates in an 8×8 pool is 4. ApproxM Pooling provides new insight into designing a novel pooling method for improving the test accuracy and weight update quantity ratio of neural networks, such as object detection, semantic segmentation and object segmentation.

References

1. Pawar, K., Egan, G.F., Chen, Z.: Domain knowledge augmentation of parallel MR image reconstruction using deep learning. Comput. Med. Imaging Graph. **92**(2), 101968 (2021)
2. Pan, Y., Jing, Y., Wu, T., Kong, X.: Knowledge-based data augmentation of small samples for oil condition prediction. Reliab. Eng. Syst. Saf. **217**, 108114 (2022)
3. He, K., Zhang, X., Ren, S., and Sun, J.: Deep residual learning for image recognition. In: Proceedings of the IEEE Conference on Computer Vision and Pattern Recognition, pp. 770–778
4. Devlin, J., Chang, M., Lee, K., Toutanova, K.: BERT: pre-training of deep bidirectional transformers for language understanding. In: Proceedings of the 2019 Conference of the North American Chapter of the Association for Computational Linguistics: Human Language Technologies, pp. 4171–4186 (2009)
5. Yann, L., Carpentier, J., Aubry, M., Sivic, J.: Single-view robot pose and joint angle estimation via render & compare. In: Proceedings of the IEEE Conference on Computer Vision and Pattern Recognition. pp. 1654–1663 (2021)
6. Gholamalinezhad, H., Khosravi, H.: Pooling methods in deep neural networks, a Review. CoRR (2020)
7. Krizhevsky, A., Sutskever, I., Hinton, G.E.: ImageNet classification with deep convolutional neural networks. In: Proceedings of International Conference on Neural Information Processing Systems, pp. 1097–1105 (2012)
8. Simonyan, K, Zisserman, A.: Very deep convolutional networks for large-scale image recognition. Comput. Sci. (2014)
9. Zagoruyko, S., Komodakis, N.: Wide residual networks. In: Proceedings of the British Machine Vision Conference, pp. 1–12
10. Cun, Y.L., Boser, B., Denker, J.S., Henderson, D., Jackel, L.D.: Handwritten digit recognition with a back-propagation network. In: Proceedings of Advances in Neural Information Processing Systems, pp. 396–404 (1999)
11. Yann, L., Bottou, L., Bengio, Y., Haffner, P.: Gradient-based learning applied to document recognition. Proc. IEEE **86**, 2278–2324 (1998)
12. Ranzato, M., Boureau, Y.L., Lecun, Y.: Sparse feature learning for deep belief networks. In: Proceedings of Advances in Neural Information Processing Systems, pp. 1185–1192 (2007)

13. Ma, T., Yang, M., Rong, H., Qian, Y., Tian, Y., Najla, A.: Dual-path CNN with Max Gated block for text-based person re-identification. Image Vision Comput. (111), 104168 (2021)
14. Shi, B., et al.: Deep panoramic representation for 3-d shape recognition. IEEE Sig. Process. Lett. **22**, 2339–2343 (2015)
15. Zeiler, M.D., Fergus, R.: Stochastic pooling for regularization of deep convolutional neural networks. preprint arXiv:1301.3557 (2013)
16. Yu, D., Wang, H., Chen, P., Wei, Z.: Mixed pooling for convolutional neural networks. In: Miao, D., Pedrycz, W., Ślęzak, D., Peters, G., Hu, Q., Wang, R. (eds.) RSKT 2014. LNCS (LNAI), vol. 8818, pp. 364–375. Springer, Cham (2014). https://doi.org/10.1007/978-3-319-11740-9_34
17. Shi, Z., Ye, Y., Wu, Y.: Rank-based pooling for deep convolutional neural networks. Neural Netw. **83**, 21–31 (2016)
18. Krizhevsky, A., Hinton, G.: Learning multiple layers of features from tiny images. Handbook of Systemic Autoimmune Diseases, vol. 1, no. 4 (2009)

Bioinformatics-Based Acquisition of Alzheimer's Disease Hub Genes

Meng-Ting Hou, Xi-Yu Li, and Juan Bao(✉)

School of Public Health, Hubei University of Medicine, Shiyan 442000, China
bao_0203@126.com

Abstract. OBJECTIVE: To analyze the functions and pathways of differentially expressed genes (DEGs) between Alzheimer's disease (AD) patients and normal controls using bioinformatics methods, and to screen Hub genes to provide theoretical support for the study of AD pathogenesis and therapeutic targets. METHODS AD-related data microarrays (GSE197505) were obtained from the GEO database, and the data were processed using GEO2R to obtain DEGs. The screened DEGs were enriched for Gene Ontology (GO) and Kyoto Encyclopedia of Genes and Genomes (KEGG) signaling pathways, using the Matascape platform. Cytoscape software was utilized to map the PPI network and screen the Hub genes. RESULTS A total of 142 DEGs were screened as down-regulated genes in this study. GO/KEGG analysis showed that these DEGs were involved in biological processes such as positive regulation of cell cycle protein-dependent serine/threonine kinase activity, and PI3K-Akt signaling pathway. Eight Hub genes were finally screened by the PPI network, four of which were validated by the literature. CONCLUSION The results of the bioinformatics network analysis revealed the Hub genes of AD, contributing to a better understanding of the mechanisms of AD and facilitating the discovery of new therapeutic targets.

Keywords: Alzheimer's Disease · Bioinformatics · Differentially Expressed Genes · Hub Genes

1 Introduction

The 2015 Global Alzheimer's Disease Report, "Dementia's Impact on the World," has reported that with the trend of population aging, China has the largest number of Alzheimer's disease patients in the world, up to 9.5 million [1]. AD is a neurodegenerative disease characterized by chronic cognitive impairment with atrophy of the cerebral cortex, deposition of β-amyloid, neuronal fibrillary tangles, and the formation of senile plaques. AD is the most common type of dementia, accounting for 50–70% of all types of dementia [2]. The World Alzheimer's Disease Report 2019, published by Alzheimer's Disease International, states that there were an estimated 50 million people globally living with dementia in 2019, and the number is projected to reach 152 million in 2050. Additionally, the cost associated with treating dementia is currently one trillion dollars per year and is projected to double by 2030 [3]. The insidious onset of AD, the slow and irreversible course of the disease, and the increasing burden of the

S. Zhang et al. (Eds.): BigData 2023, LNCS 14203, pp. 119–127, 2023.
https://doi.org/10.1007/978-3-031-44725-9_9

disease pose serious challenges to the prevention and treatment of AD. Regarding the diagnosis of AD, the more established markers are amyloid and Tau proteins, and the presence of markers in the cerebrospinal fluid is indicative of a patient with AD. Genetic data can also provide an important basis for diagnostic aid for researchers. It has been found that the apolipoprotein E4 gene and the clumping protein gene can control the process of amyloid clearance, and the lack of these genes will result in AD disease [4]. In this study, bioinformatics methods were utilized to identify and screen Hub genes that may be involved in the occurrence and development of AD, to explore the mechanism of action of these genes, and to provide new targets and a basis for the diagnosis and treatment of AD.

2 Tools and Methods

2.1 Tools

GEO Database and GEO2R. The GEO Database is a public repository for storing and sharing gene expression data and is the most comprehensive public gene expression database available today [5]. GEO2R is an online tool available in the GEO database that analyzes differentially expressed genes using the R language.

STRING and Cytoscape. The STRING database integrates protein information data from multiple sources to create a comprehensive protein interaction network. Cytoscape is an open source software platform for visualizing molecular interaction networks and biological pathways.

GO and KEGG. GO is a standardized classification system widely used in gene function annotation and bioinformatics research. KEGG is a comprehensive bioinformatics database and resource that provides information on gene signaling pathways and more.

Metascape. Metascape is an online bioinformatics analysis platform that integrates more than 40 bioinformatics knowledge bases, including GO, KEGG, UniProt and DrugBank, enabling pathway enrichment as well as bioprocess annotation.

2.2 Methods

Dataset Download Source and Extraction. Enter the search keyword Alzheimer's disease in the GEO database with the search formula: ("alzheimer disease"[MeSH Terms] OR ("alzheimer disease"[MeSH Terms] OR Alzheimer's disease[All Fields])) AND "Homo sapiens"[organism] AND "NORMAL", select SERIES. Download gene chip GSE197505 from the GEO database.

Differential Expression Genes Screening. The t-test was performed using GEO2R on the genes of AD patients' brain tissues and normal brain tissues from GSE197505 gene chips, and the DGEs were screened out by statistical methods. Define $\log_2 FC > 1$ as up-regulated differentially expressed gene "UP" and $\log_2 FC < -1$ as down-regulated differentially expressed gene "DOWN".

GO and KEGG Signaling Pathway Enrichment Analysis. Enter the Gene ID of the DEGs in Metascape, select Homo sapiens for the sample type, select Custom Analysis for the analysis method, and then perform GO and KEGG enrichment analysis.

Screening for Hub Genes. In the STRING database, select multi-protein analysis, enter the gene ID of the DEGs, and select Homo sapiens for the sample type to obtain the key gene PPI network of the differentially expressed gene. Download the relevant data into Cytoscape software, visualize the data, and then use the Mcode plug-in to screen and map the PPI network of Hub genes.

3 Results

3.1 GeneChip Characterization

The AD GeneChip GSE197505 selected for this study contains frontal cortical isolated extracellular vesicles from 8 AD patient donors and 10 normal control frontal cortical isolated extracellular vesicles (EVs) [6]. EVs are carriers of nucleic acids, lipids and proteins and play an important role in the pathogenesis of neurodegenerative diseases.

3.2 Differentially Expressed Genes Screening

Differentially expressed genes were screened by GEO2R on GeneChip GSE197505, with screening thresholds set: $|\log_2 FC| > 1$ and $P.adj < 0.05$, and after probe de-duplication, a total of 142 DEGs were read, all of which were down-regulated genes. A volcano map of differentially expressed genes was drawn (see Fig. 1). The multidimensional data of this data set was down-analyzed (see Fig. 2), and using the Uniform Manifold Approximation and Projection (UMAP, consistent popular approximation and projection) method, it can be seen that the data set was separated into high expression group and ground expression group.

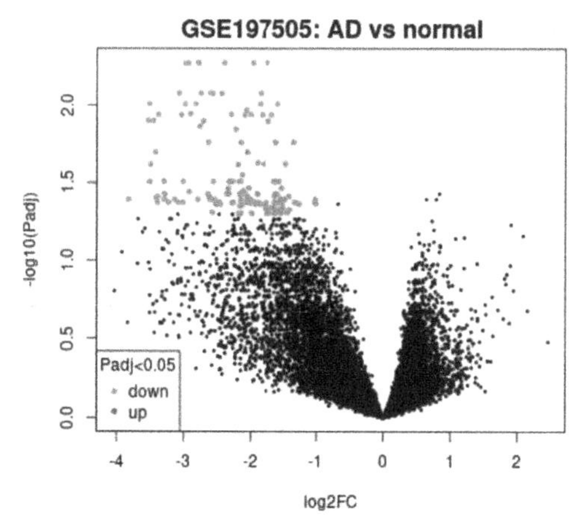

Fig. 1. Volcano plot of differentially expressed genes. This figure shows that the differentially expressed genes are all down-regulated genes (blue dots are down-regulated genes and black dots are genes with insignificant differential expression). (Color figure online)

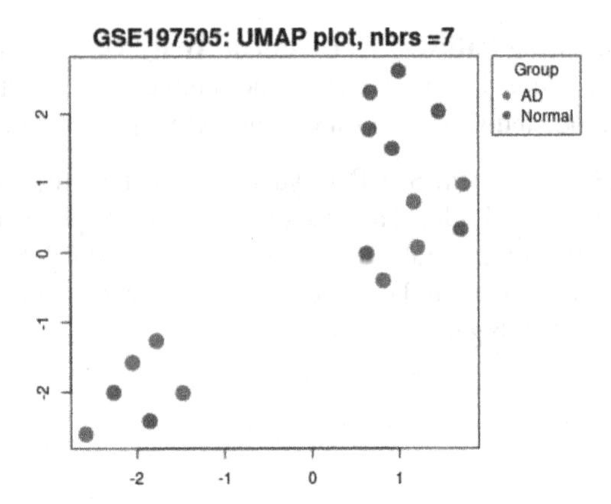

Fig. 2. UMAP diagram of GSE197505.

3.3 Functional Analysis of Differentially Expressed Genes

Metascaped was used to annotate the differentially expressed genes with gene ontology markers and related metabolic pathways and biological processes. GO enrichment analysis showed that the molecular functions of the differentially expressed genes in the GeneChip GES197505 mainly included cytokine binding, calcium binding, antigen binding, etc., with cellular sublocalization mainly in the cellular matrix, and involved in biological processes mainly in positive regulation of The differentially expressed genes were mainly involved in the positive regulation of cell cycle protein-dependent serine/threonine kinase activity, vascular morphology, response to organophosphorus, and peptide lysine modification, etc. (see Fig. 3). KEGG enrichment analysis showed that the major enrichment pathways of the differentially expressed genes included oocyte meiosis, PI3K-Akt signaling pathway, and the synthesis of thyroid hormones (see Fig. 4). Among them, the PI3K-Akt pathway plays an important role in the survival, proliferation and functional maintenance of neuronal cells. Abnormal activation or impairment of this pathway is associated with neurological disorders such as Alzheimer's disease, Parkinson's disease, and neurodegenerative diseases [7].

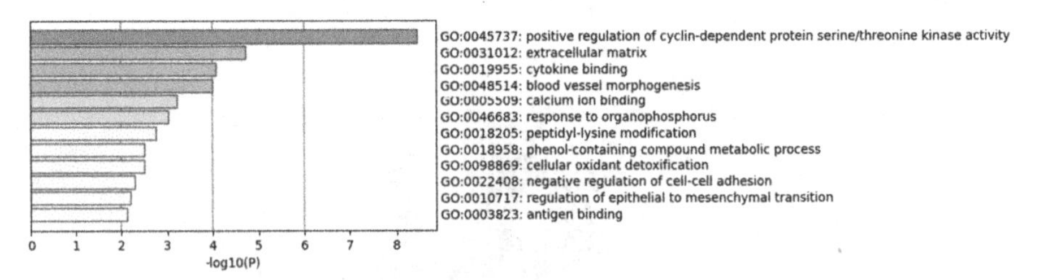

Fig. 3. Bar graph of GO enrichment analysis of differentially expressed genes.

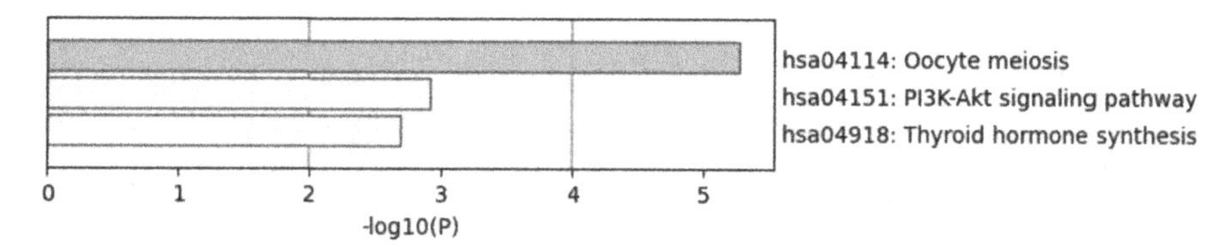

Fig. 4. Bar graph of KEGG pathway enrichment analysis of differentially expressed genes.

3.4 Relationship of Hub Genes in Protein-Protein Interaction Networks

The data related to protein-protein interaction information were obtained from the STRING database and imported into Cytoscape to analyze and visualize to construct a PPI network (see Fig. 5). This PPI network has 89 nodes and 176 edges. By KEGG analysis, these key genes were mainly closely related to progesterone-mediated oocyte maturation and oocyte meiosis. Eight Hub genes that play a pivotal role in this PPI network were screened using the MCODE plug-in (see Fig. 6), namely, LRRC32, ITPR3, SLC26A6, NOS3, ADAM33, PIEZO1, GSDMB, IL4R.

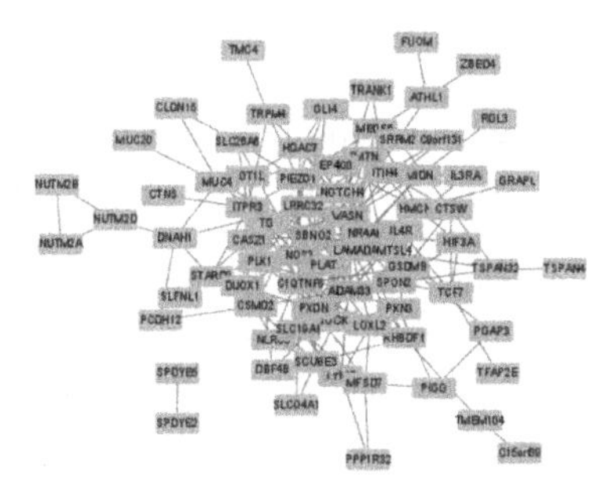

Fig. 5. PPI network of differentially expressed genes.

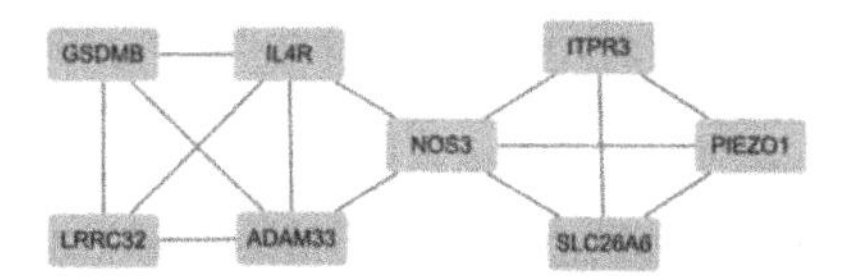

Fig. 6. Hub genes in the PPI network.

4 Discussions

4.1 Significance of Hub Genes for Alzheimer's Disease

With its insidious onset, slow and irreversible course, and increasing disease burden, AD has become one of the greatest global public health and social challenges facing humanity today and in the future [8]. In this study, combining with GEO, STRING and other related databases, we screened 8 Hub genes related to AD (LRRC32, ITPR3, SLC26A6, NOS3, ADAM33, PIEZO1, GSDMB, IL4R) using bioinformatics technology by annotating the gene functions, molecular metabolic pathways, and the establishment of PPI networks. Finally, the relationship between Hub genes and AD is analyzed in the light of domestic and international studies and literature, which will provide new ideas for the study of the pathogenesis of AD and help in the early diagnosis and treatment of AD.

4.2 Mechanisms of Hub Genes Action in AD

NOS3. NOS3 belongs to the nitric oxide synthase family, and in 1999, Dahiyat [9] conducted a study investigating the association between the NOS3 gene and AD, suggesting that changes in the nitric oxide synthase system may influence AD-related pathogenesis, and that NOS3 gene expression induces neuronal and glial degeneration in the brain. Superoxide free radicals react with NO to produce peroxynitrite, which promotes lipid peroxidation, which in turn further accelerates degenerative changes and causes AD [10]. The arginine metabolic pathway produces the gaseous signaling molecule NO mainly through NOS, and the NO produced by eNOS is essential for maintaining normal cerebral blood flow [11], and previous studies have verified that the formation of plaques and tangles in the AD brain is closely related to the reduction of capillary eNOS expression [12, 13]. NO from nNOS has been found to play an important role in synaptic plasticity and learning memory, while NO from iNOS is pro-inflammatory [14]. It is directly associated with 1- glutamate and NMDA receptors in the CNS, and therefore with long time-range potentiation (LTP), which is considered a major cellular mechanism of learning and memory, and synaptic transmission is enhanced by repeated stimulation of presynaptic terminals. Therefore, large amounts of Ca2 + are expressed through NMDA receptors located on the postsynaptic membrane, and calcium/calmodulin-mediated regulation of nNOS/NO shows a potential induction of LTP [15]. The results of GO enrichment analysis in this study also showed that the molecular functions of the differentially expressed genes in GeneChip GES197505 included calcium ion binding. Further studies on the mechanism of action of NOS3 are still needed.

Piezo1. Alzheimer's disease is associated with beta-amyloid deposition (Aβ). The mechanosensitive ion channel Piezo1 is increased in microglia in response to stiff stimulation of Aβ fibers. Upregulation of Piezo1 in Aβ plaque-associated microglia was observed in AD mouse models and human patients. Piezo1-deficient microglia interfered with microglia aggregation, phagocytosis, and Aβ plaque compaction, leading to exacerbation of Aβ and neurodegenerative lesions in AD. In contrast, activation of Piezo1 ameliorated cerebral Aβ load and cognitive deficits in 5xFAD mice [16]. Another study demonstrated that the mechanosensitive cation channel Piezo1 plays a key role in translating ultrasound-related mechanical stimuli through its trimeric propeller-like structure, but the importance of Piezo1-mediated mechanotransduction in brain function has not been sufficiently emphasized. In addition to mechanical stimulation, Piezo1 channels are strongly modulated by voltage. Fangxuan C et al. [17] hypothesized that Piezo1 may play a role in the conversion of mechanical and electrical signals to induce phagocytosis and degradation of Aβ and β-phosphodiester. To this end, a transcranial magnetoacoustic stimulation (TMAS) system was designed and applied to 5xFAD mice to assess whether TMAS could alleviate the symptoms of AD mouse models by activating Piezo1. Finally, it was confirmed that Piezo1 can convert TMAS-related mechanical and electrical stimuli into biochemical signals, and it was determined that the favorable effects of TMAS on synaptic plasticity in 5xFAD mice are mediated by Piezo1. This could lead to the use of Piezo1 as a candidate target for AD therapy and provide new ideas for AD treatment.

GSDMB. GSDMB is a member of the GSDM protein family. Zhou et al. [18] found that granzyme A cleaved GSDMB, releasing the GSDMB-N fragment, which caused perforation of the cell membrane and led to cellular juxtaposition in GSDMB-expressing cancer cells. To explore the role of GSDMB in nonclassical juxtaposition, Chen et al. [19] specifically knocked down GSDMB in THP-1 cells and found that the induced nonclassical juxtaposition was inhibited. In contrast, the high expression of GSDMB induced using lentiviral system showed promotion of nonclassical juxtaposition. It was verified that GSDMB was involved in nonclassical juxtaposition. In addition, the team found that GSDMB promoted the enzymatic activity of caspase-4 by combining with caspase-4, which led to the cleavage of GSDMD by caspase-4, thus releasing GSDMD-N-terminal protein leading to cellular juxtaposition, which led to the release of inflammatory factors in large quantities, resulting in inflammatory diseases. Microglia focal death caused by neuroinflammatory response is closely related to the pathogenesis of many neurological pathologies (e.g., stroke, depression, neurodegenerative diseases, etc.), and it is the main cell type in which focal death of cells occurs [20]. Liang et al. [21] found that microglia were able to secrete ASCs under the action of Aβ fibers, a major neurotoxic component of Alzheimer's disease patients, and to produce NLRP3-dependent IL-1β inflammatory factor, which induces cellular pyroptosis and further amplifies the inflammatory response; thus confirming that the deletion of the ASC gene and the blockade of antibodies play an inhibitory role in the recruitment of Aβ fibers, suggesting that inhibition of cellular pyroptosis-released ASC may slow down the progression of Alzheimer's disease. The team also found that the brains of AD model mice highly expressed the cellular focal death-related molecules NLRP3 inflammatory vesicles and Caspase-1, and that deletion of the NLRP3 or Caspase-1 genes greatly improved spatial memory and enhanced Aβ clearance in the mice. These studies reveal novel mechanisms of GSDMB-mediated nonclassical cellular focal death associated with AD, providing potential therapeutic strategies and targets.

ADAM33. The ADAM33 gene is localized on the short arm of chromosome 20, 20p13, and belongs to the ADAM gene superfamily, which consists of eight structural domains. The gene is widely expressed in the human body, but is highly expressed in tissues such as the brain and lungs. The proteins encoded by the ADAM gene superfamily have metalloprotease activity, and this activity has been associated with the development of a variety of diseases [22]. The protease activity of ADAM has a protective function for the organism. For example, ADAM9, −10, and −17 have α-secretase activity, and in patients with Alzheimer's disease, α-secretase cleaves amyloid precursor protein (APP) from the middle of amyloid β peptide (Aβ), attenuates the deposition of β-peptide amyloid plaques and promotes the release of neuroprotective agent, sAPPα. Therefore, the α-secretase activity of ADAM molecules provides a a new tool for the treatment of Alzheimer's disease [23]. The ADAM33 gene can be a candidate gene for asthma, and its inhibitors are expected to be used in the treatment of asthma [24]. And further studies on the relationship between ADAM33 and AD are needed.

In summary, PIZO1 and GSDMB may provide targets for the treatment of AD, NOS3 and ADAM33 can be further investigated as candidate genes, whereas IL4R,Lrrc32,ITPR3, and SLC26A6 have been less studied in AD, and their mechanisms of action need to be further investigated.

5 Conclusion

This study utilized a series of bioinformatics methods to screen multiple Hub genes, some of which have been shown to serve as therapeutic targets for AD. The newly screened Hub genes will provide ideas and rationale for understanding the pathogenesis of AD and exploring new therapeutic options.

Acknowledgements. This work was supported by the Start-up Foundation of Hubei University of Medicine (No. 2019QDJRW02); Cooperative Education Program of the Ministry of Education(No.202101142004); Medical Education Research Project of Chinese Medical Association(No.2020B-N02353).

References

1. Prince, M.: The Global Impact of Dementia an analysis of prevalence, incidence, cost and trends. World Alzheimer Report 2015 (2015)
2. China dementia and cognitive disorders guideline writing group.: cognitive disorder disease specialized committee of neurologists branch of Chinese physicians association. 2018 China dementia and cognitive disorders diagnostic and treatment guidelines (I): diagnostic criteria for dementia and its classification. Chin. Med. J. **98**(13), 965–970 (2018)
3. Alzheimer's Disease International.: World Alzheimer Report 2019: Attitudes to dementia. Alzheimer's Disease International, London (2019)
4. Ruoqi, Z.: Progress in molecular imaging of pathological markers of Alzheimer's disease. Shanxi Med. J. **50**(3), 381–384 (2021)
5. Barrett, T.: NCBI GEO: archive for high-throughput functional genomic data. Nucleic Acids Res. **37**(Database issue), D885-D890 (2019)
6. Dan, L.: Long RNA profiles of human brain extracellular vesicles provide new insights into the pathogenesis of Alzheimer's disease. Aging Dis. **14**(1), 168–178 (2023). PMID: 36818567
7. Manish, K.: Implications of phosphoinositide 3-kinase-akt (PI3K-Akt) pathway in the pathogenesis of Alzheimer's disease. Mol. Neurobiol. **59**(1), 52–64 (2021)
8. Blennow, K.: Evolution of Abeta42 and Abeta40 levels and Abeta42/Abeta40 ratio in plasma during progression of Alzheimer's disease: a multicenter assessment. J. Nutr. Health Aging **13**(3), 232–236 (2009)
9. Dahiyat, M.: Association between Alzheimer's disease and the NOS3 gene. Ann. Neurol. **46**(4), 664–667 (1999)
10. Carmen, G.: Increased susceptibility to plasma lipid peroxidation in Alzheimer disease patients. Curr. Alzheimer Res. **1**(2), 103–109 (2004)
11. Bergin, D.: Altered plasma arginine metabolome precedes behavioural and brain argi nine metabolomic profile changes in the APPswe/PS1ΔE9 mouse model of Alzheimer's disease. Transl. Psychiatry **8**(1), 108 (2018)

12. Jeynes, B.: Significant negative correlations between capillary expressed eNOS and Alzheimer lesion burden. Neurosci. Lett. **463**(3), 244–248 (2009)
13. John, P.: Neurofibrillary tangles and senile plaques in Alzheimer's brains are associated with reduced capillary expression of vascular endothelial growth factor and endothelial nitric oxide synthase. Curr. Neurovasc. Res. **5**(3), 199–205 (2008)
14. Robert, F.: NO/cGMP-dependent modulation of synaptic transmission. Handb. Exp. Pharmacol. **184**, 529–560 (2008)
15. Harikesh, D.: Alzheimer's disease: a contextual link with nitric oxide synthase. Curr. Mol. Med. **20**(7), 505–515 (2020)
16. Jin, H.: Microglial piezo1 senses Aβ fibril stiffness to restrict Alzheimer's disease. Neuron **99**(1), 122–135 (2022)
17. Fangxuan, C.: Transcranial magneto-acoustic stimulation attenuates synaptic plasticity impairment through the activation of piezo1 in Alzheimer's disease mouse model. Research **6**(3), 58–67 (2023)
18. Zhiwei, Z.: Granzyme A from cytotoxic lymphocytes cleaves GSDMB to trigger pyroptosis in target cells. Science **368**(6494), 1260–1263 (2020)
19. Chen, Q.: Molecular mechanism of GSDMB enhancing the enzymatic activity of cysteoaspartase caspase-4 and thus promoting nonclassical cellular pyrokinesis. Nanjing University (2018)
20. Chunkai, J.: Progress of cellular focal death in stroke and depression. J. Stroke Neurol. Dis. **40**(05), 463–467 (2023)
21. Fei, L.: High-intensity interval training and moderate-intensity continuous training alleviate β-amyloid deposition by inhibiting NLRP3 inflammasome activation in APPswe/PS1dE9 mice. NeuroReport **31**(5), 425–432 (2020)
22. Primakoff, P.: The ADAM gene family: surface proteins with adhesion and protease activity. Trends Genet. **16**(2), 83–87 (2000)
23. Masashi, A.: Putative function of ADAM9, ADAM10, and ADAM17 as APP alpha-secretase. Biochem. Biophys. Res. Commun. **301**(1), 231–235 (2003)
24. Zhiguang, S.: Progress in the study of ADAM33 gene. Genetics **04**, 636–640 (2005)

Application Track

Enhanced Campus Information Query System based on ChatGPT Interface and Local Content Database

Kang Minjie[1], Ji Ran[1], Gui Ao[1], Pang Xuejiao[1], Fan Xiaohu[1,2,3](✉), Yi Li[4], Lu Xing[1], and Han Jie[1]

[1] Wuhan College, Wuhan 430070, Hubei, China
{9452,9420,8201,8208}@whxy.edu.cn, {20202140503, 20203170218}@smailwhxy.edu.cn
[2] Wuhan Tuspark Hezhong Science and Technology Develop Co. Ltd, Wuhan 430070, China
[3] Wuhan Bohu Science and Technology Co. Ltd, Wuhan 430074, China
[4] Shenzhen Institute of Information Technology, Shenzhen 518000, China
20202140527@smailwhxy.edu.cn

Abstract. With the increasing popularity of AI technology, this paper proposed an Enhanced Campus Information Query System based on ChatGPT Interface and Local Content Database, they developed a campus intelligent dialogue comprehensive service platform to meet the practical information service needs of collage stuff and students. The platform utilizes the latest ChatGPT model API for secondary development. Node.js technology is used as the backend, combined with the ChatGPT model API to achieve natural language interaction. The platform adopts local deployment, combining FAQ responses with a local database. By applying Dynamic Programming algorithm and Levenshtein distance algorithm, the platform implements keyword matching and fuzzy query functions. The dynamic programming algorithm is used to calculate the similarity score of strings by comparing the similarity between two strings and giving a numerical score. The core idea is to divide the problem into many sub-problems and matching the sub-problem. In keyword matching, dynamic programming algorithm can be used to calculate the similarity score between user input and keywords to determine the best match. At the same time, fragmented internal campus information resources are integrated, and users can obtain relevant information through keyword matching queries and enjoy personalized services and recommendations. The platform supports interactive dialogue form, making it convenient for users to quickly obtain the required information. In addition, our algorithm has significantly improved accuracy in keyword matching and fuzzy queries, increasing from 80% to 95%, and efficiency has increased by 50%. Moreover, the new algorithm can handle longer and more complex query strings and more query conditions, meeting the complex query needs of users. Convenient services are provided through universal and user-friendly methods such as WeChat Mini Program and web, improving user experience and satisfaction.

Keywords: Campus Information Interaction System · ChatGPT · Dynamic Programming · Levenshtein Distance

S. Zhang et al. (Eds.): BigData 2023, LNCS 14203, pp. 131–148, 2023.
https://doi.org/10.1007/978-3-031-44725-9_10

1 Introduction

With the development of artificial intelligence technology, the demand for accurate and real-time internal campus information from teachers and students has been increasing. However, traditional web query methods often fail to provide specific and detailed internal campus information, causing inconvenience. Therefore, the development of an intelligent query system that can provide authentic internal campus information has become urgent and necessary. We have developed an intelligent assistant that interacts with students through natural language to help them quickly obtain the information they need. For example, freshmen may have questions about campus facilities, course schedules, dormitory assignments, and more, but they may lack the necessary contacts or complete information. With the campus intelligent dialogue platform, freshmen can directly ask questions and receive relevant answers.

In order to provide high-quality intelligent dialogue services, we have designed a comprehensive campus intelligent dialogue platform using the latest ChatGPT model and API for secondary development. We have also employed a combination algorithm of Node.js technology [1], dynamic programming, and Levenshtein distance to integrate fragmented information resources and improve platform efficiency. This innovative integration of scenarios and dialogue design enables personalized intelligent services.

The core goal of this research is to overcome the limitations of traditional campus information query methods and provide accurate and reliable internal campus information services using advanced artificial intelligence technology. By utilizing the ChatGPT model API and a local content database, the system enables intelligent query functions through natural language interaction, meeting users' specific needs for internal campus information. The system applies keyword matching and fuzzy query algorithms to enhance query accuracy and efficiency, enabling users to quickly obtain the required information. Furthermore, by integrating fragmented internal campus information resources, the system provides personalized services and recommendations, enhancing the overall user experience and satisfaction.

2 Related works

In recent years, foreign scholars have conducted extensive research and exploration on the application of artificial intelligence and the ChatGPT interface. Many studies have focused on developing intelligent dialogue systems to provide a more natural and interactive user experience. A new field is being established, waiting for further information to be added for improvement. Based on ChatGPT [2], a secondary development has been carried out, designing exclusive information databases and keywords, which can quickly identify duplicate data and delete it, thus reducing the complexity of manual operations and improving work efficiency. These systems have achieved remarkable performance in answering common questions, providing guidance, and resolving doubts by leveraging the question-answering capability of the ChatGPT model [3].

Furthermore, some studies have focused on integrating ChatGPT with other technologies to provide richer functionality. For example, by combining knowledge graphs or domain expert systems with the ChatGPT model, knowledge acquisition, reasoning, and problem-solving can be accomplished [4]. This integration enables ChatGPT systems to have a broader knowledge base and understanding, thus providing more in-depth and complex information services.

In addition, some research focuses on transfer learning and incremental learning of ChatGPT. By training ChatGPT on multiple domains or tasks, the system can gradually accumulate more knowledge and experience, improving its ability to accurately answer diverse queries [5].

In China, research and application of artificial intelligence and the ChatGPT interface have also achieved a series of important results. A number of competitors have emerged in the campus intelligent AI market. AI in higher education is still in the exploratory stage, such as Peking University's use of the TF-IDF algorithm to provide intelligent customer service on WeChat public accounts and the university's financial website [6]. Yan Shuo et al. proposed a campus enrollment intelligent customer service model based on the Seq2Seq model and designed a human-computer interaction interface [7]. These efforts aim to improve the quality and efficiency of services. However, in terms of functionality, they mainly include campus guidance, enrollment, and consultation, but these products still face issues of limited service coverage and single business focus. Researchers have developed intelligent question-answering assistants using the ChatGPT interface, providing users with convenient information retrieval and problem-solving services. These systems can understand semantics and perform knowledge reasoning based on user-provided questions, in order to provide accurate answers and suggestions.

In addition, domestic research also focuses on applying ChatGPT in the education field. Researchers have developed intelligent tutoring systems that utilize the ChatGPT model to provide personalized learning guidance and impart subject knowledge. Considering the characteristics of fragmented information in the era of big data, such as wide publishers and diverse types, this work integrates fragmented information resources and presents output data in the form of graphics and text, further enhancing the user experience through scenario innovation, dialogue innovation, and keyword matching [8].

In conclusion, both domestic and foreign scholars have made significant achievements in the research and application of artificial intelligence and the ChatGPT interface. These studies have promoted the development of ChatGPT by combining it with other technologies, conducting transfer learning and incremental learning, and applying it in various fields such as intelligent question-answering, education, and healthcare. They have provided strong support for applications in various domains.

3 System Architecture

3.1 Top-Level Architecture Design

Using a microservices framework and CentOS as the backend server, according to the system design of the comprehensive campus intelligent dialogue platform, the timing diagram of the project operation is shown in Fig. 1.

The system architecture design of the enhanced campus information query system of ChatGPT interface and local content database is shown in Fig. 2. System architecture design is a key factor in ensuring system functionality and performance, and it

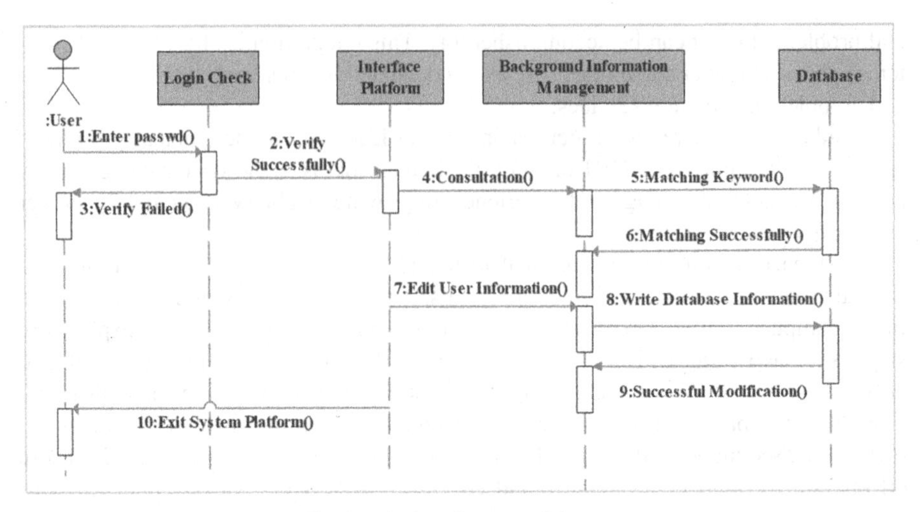

Fig. 1. Timing diagram of System

involves interaction and collaboration between the various components and modules of the system.

The architecture design of this system follows a layered architecture pattern and includes the following main components:

Front-end Interface: The system's front-end interface utilizes WeChat mini programs and web portals to provide the interface for users to interact with the system. Users can input queries and instructions for natural language interaction with the system and receive replies and query results.

Back-end Processing: The system's back-end utilizes Node.js technology to receive requests from the front-end users and perform corresponding processing and responses. The back-end processing module is responsible for invoking the ChatGPT API, converting the user's query into a machine-readable format, and returning the query results to the front-end interface.

ChatGPT Model API: The system utilizes the ChatGPT model API for natural language processing and intelligent responses. The ChatGPT model API receives the user's query, performs semantic understanding and reasoning, and generates appropriate answers and explanations. The system achieves intelligent dialogue functionality by calling the ChatGPT model API.

Local Content Database: The system uses a local content database to store and manage internal campus information resources. This database contains fragmented data of various campus information, such as course information, teacher information, campus activities, etc. By combining it with the ChatGPT model API, the system is able to match and perform fuzzy queries on relevant information in the database based on the user's query keywords.

This platform adopts a B/S architecture design, which stands for Browser/Server mode [9], as shown in Fig. 3.

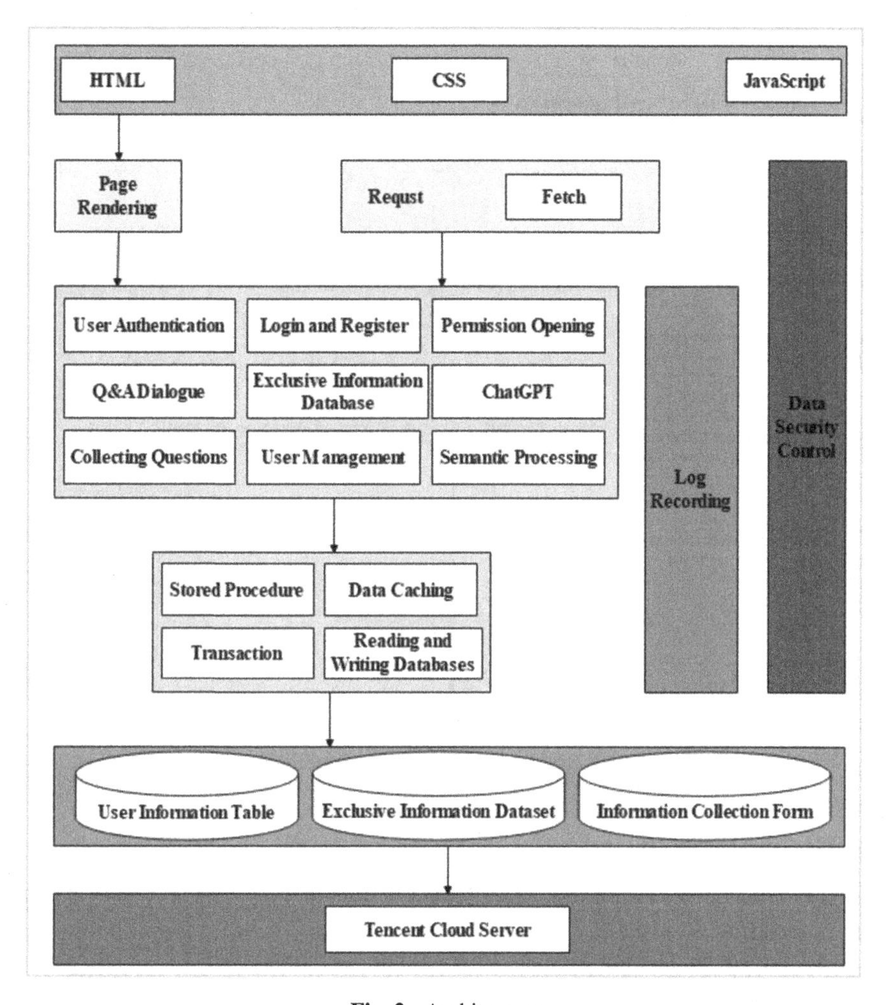

Fig. 2. Architecture

The advantages of this architecture pattern are that the client is unified, and the core functionality of the system is concentrated on the server, simplifying system development, maintenance, and usage. Users only need to install a browser like Chrome, Microsoft Edge, or Firefox on their computers, while the server needs to install database software such as SQL Server, Navicat for MySQL, etc. Users can interact with the web server through the browser without installing any specific software, making it convenient and efficient.

Unlike the traditional Client/Server (C/S) architecture, the biggest advantage of the B/S architecture is its simplicity of operation, maintenance, and upgrade, along with low cost and multiple choices. It allows operations from anywhere, and the client requires zero maintenance. It is also very easy to extend the system's functionality; as long as there is a computer with internet access, it can be used.

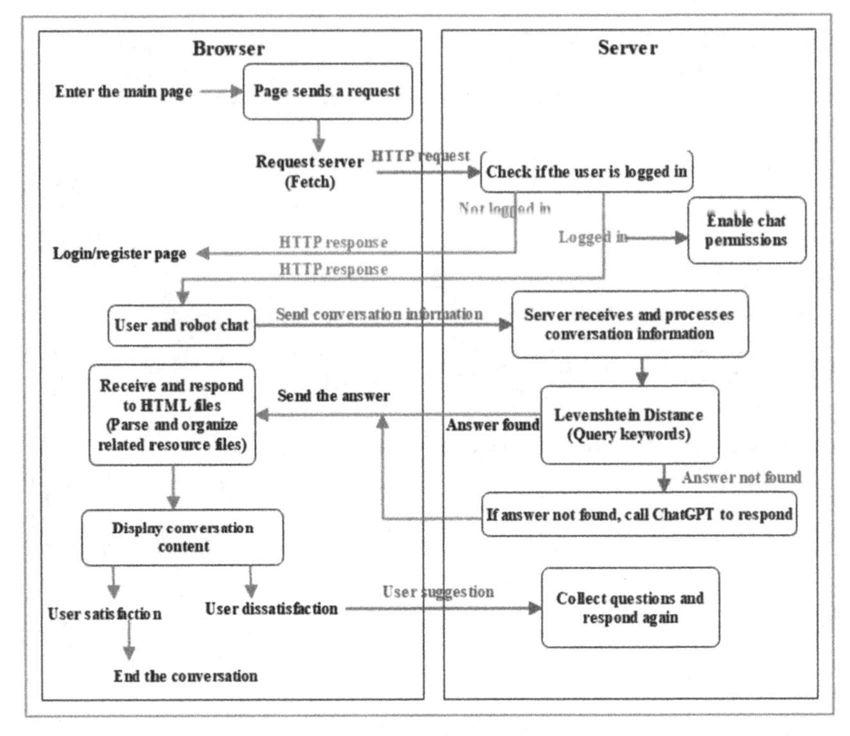

Fig. 3. B/S structure sending and responding process

3.2 Module Hierarchy

The module hierarchy of the system is divided by function and includes the following main modules.

Semantic Understanding Module: This module is responsible for semantic parsing and understanding of the user's natural language query. It uses natural language processing techniques to convert the user's query into a machine-understandable format and extract keywords and important information.

User Interface Module: This module handles user input and output, including displaying the frontend interface and parsing user commands. Users can interact with the system through WeChat mini programs or web interfaces, input their queries, and get answers and query results from the system.

ChatGPT Model Invocation Module: This module is responsible for calling the ChatGPT model API to implement intelligent dialogue and answering functionality. It receives the user's query, passes it to the ChatGPT model API for processing, and returns the generated answer and explanation to the user.

Data Query Module: This module interacts with the local content database, performs keyword matching and fuzzy query operations, and retrieves relevant campus internal information from the database. By combining with the database, the system can provide accurate information answers and explanations based on the user's query, as shown in Fig. 4.

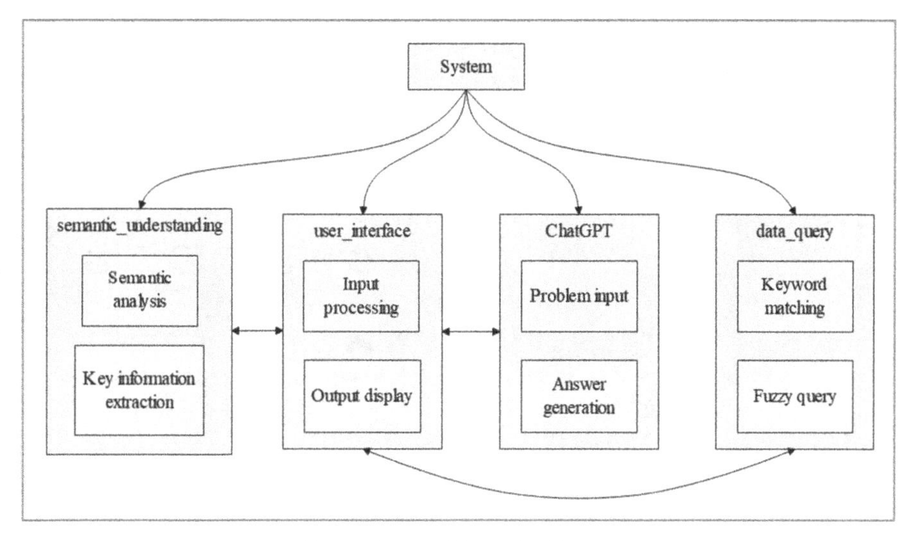

Fig. 4. Module hierarchy diagram

3.3 Interface Design

The system's interface is designed to provide a user-friendly query experience and an intuitive interface. Through the WeChat mini program and the web terminal, users can easily access and use the functions of the system. This is shown in Fig. 5 and Fig. 6.

Interface Design Elements include the following aspects:

User Input Area: Users can enter query questions and instructions in the input area to interact with the system using natural language.

Query Result Display Area: Query results will be displayed to the user in a designated area of the interface. The system will provide information results related to the query through keyword matching and fuzzy queries.

Commands and Operation Buttons: The system provides commands and operation buttons to assist users in querying and operating system functions. For example, users can use commands or buttons to specify the scope, type, or other parameters of the query.

3.4 Database Design

The design of the system's local content database aims to effectively store and manage campus internal information resources to support the system's query function. The elements of the database design include the following aspects:

Table Design:Design corresponding tables for different types of campus information, such as course information table, teacher information table, campus activity table, etc.

Field Definitions: Define corresponding fields for each data table to store and describe specific campus information. Fields can include course names, teacher names, activity dates, etc.

Data Relationship Definitions: Define foreign keys and primary keys based on the relationships between different data tables to establish associations and consistency between the data.

Fig. 5. AI chat interface

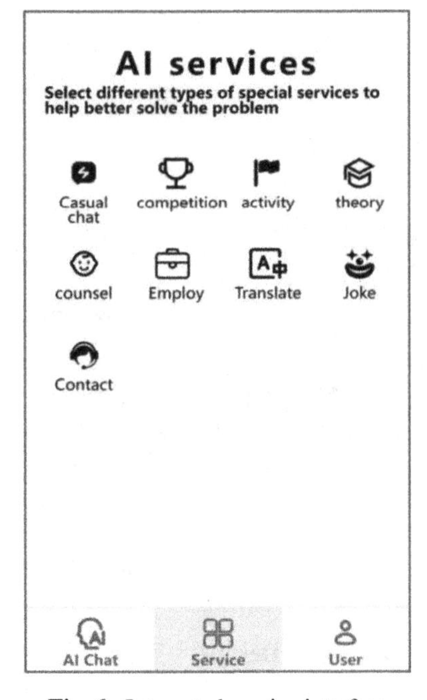

Fig. 6. Integrated service interface

Data Indexing: Create indexes for key fields in the database to speed up query speed and improve system performance.

Table 1. Userinfo table

Fields	Type	Instructions	Uniqueness	Null or not
useraccount	varchar(20)	User account	√	NO
password	varchar(20)	User password		NO
username	varchar(20)	username		NO
userheadportrait	varchar(200)	User profile picture		NO

Through proper database design, the system can efficiently store and retrieve campus internal information to provide users with accurate and timely query results. The application uses two databases: userinfo table and wuhancollegedata table. The userinfo table is used to store and access user data. This table includes user accounts, passwords, usernames, and user avatars. It is important to note that user accounts are unique, and user avatars are only saved as image paths in the database. The specific table structure is shown in Table 1.

Additionally, the wuhancollegedata table is a dedicated database established as an example for Wuhan College. It mainly consists of keywords and their corresponding

Table 2. Wuhancollegedata table

Fields	Type	Instructions	Uniqueness	Null or not
keyword	varchar(30)	keyword	√	NO
answer	varchar(500)	reply		NO

answers. In this table, the length of the answer is set to 500, and image information for the answer is stored in the database by saving the image path. The specific table structure is shown in Table 2.

4 Algorithm combination

4.1 Dynamic Programming Algorithm

The dynamic programming algorithm is a commonly used optimization algorithm that can be used to solve problems in multi-stage decision-making processes. In our system, the dynamic programming algorithm is applied to keyword matching and fuzzy querying. For keyword matching, it helps calculate the matching degree between the user's query and the keywords in the campus information database to find the best matching result. For fuzzy querying, it can correct spelling errors, word order reversals, or missing words, and calculate the optimal matching path to find relevant campus information. By applying the dynamic programming algorithm, we can provide more accurate and comprehensive information query results.

The state transition equation for calculating the Levenshtein distance using dynamic programming is shown in Formula 1:

$$d_{[i][j]}\begin{cases} 0, i = 0, j = 0 \\ i, j = 0, i > 0 \\ j, i = 0, j > 0 \\ d_{[i-1][j-1]}, S_{1[i]} = S_{2[j]} \\ \min\left(d_{[i-1][j]} + 1, d_{[i][j-1]} + 1, d_{[i-1][j-1]} + 1, S_{1[i]} \neq S_{2[j]}\right. \end{cases} \tag{1}$$

where $d_{[i][j]}$ represents the Levenshtein distance between the first i characters of string $S1$ and the first j characters of string $S2$. Specifically, when $S_{1[i]}=S_{2[j]}$, it means that the i-th character matches the j-th character, and $d_{[i][j]}$ can be derived from $d_{[i-1][j-1]}$ When $S_{1[i]} \neq S_{2[j]}$, it means that the i-th character does not match the j-th character. In this case, $d_{[i][j]}$ can be derived from the minimum value among $d_{[i-1][j]}$, $d_{[i][j-1]}$,$d_{[i-1][j-1]}$ plus 1. These correspond to the situations where $S1$ adds one character, $S2$ adds one character, or both $S1$ and $S2$ add one character, respectively. Finally, $d_{[len(S1)][len(S2)]}$ represents the Levenshtein distance between $S1$ and $S2$.

The implementation of dynamic programming algorithm to calculate Levinstein distance is as follows:

Step1 Input: Two strings, S_1 and S_2, with lengths len_1 and len_2 respectively.

Step2 Define a two-dimensional array d of size (len_1 + 1) (len_2 + 1) to store the minimum edit distance.
Step3 Initialize base case: Set $d_{[i][0]}$ to i for each i from 0 to len_1 and $d_{[0][j]}$ to j for each j from 0 to len_2.
Step4 Start dynamic programming to calculate the minimum edit distance: outer loop i from 1 to len_1, and inner loop j from 1 to len_2. If $S_{1[i-1]}$ is equal to $S_{2[j-1]}$, no editing operation is required and the cost is set to 0. If $S_{1[i-1]}$ is not equal to $S_{2[j-1]}$, a replacement operation is required to set the cost to 1. Calculate $d_{[i][j]}$, select the minimum edit distance: the minimum value in the delete ($d_{[i-1][j]}$ + 1), insert ($d_{[i][j-1]}$ + 1) and replace ($d_{[i-1][j-1]}$ + cost), and assign the result to $d_{[i][j]}$.
Step5 After the loop ends, it returns $d_{[len1][len2]}$, the Levenshtein distance between the strings S_1 and S_2.

4.2 Levenstein Distance Algorithm

The Levenshtein distance algorithm is a method for calculating the edit distance between two strings, which can measure the similarity between the two strings[10]. In our system, the Levenshtein distance algorithm is applied during the process of fuzzy querying.

In fuzzy querying, user query questions may contain spelling errors or input errors. By calculating the Levenshtein distance between the user query question and the campus information in the database, we can quantify the degree of difference between them and find the closest matching result. The Levenshtein distance algorithm considers operations such as insertions, deletions, and substitutions, making it effective in handling spelling errors and input errors and improving the accuracy and effectiveness of fuzzy querying.

For two strings A and B, the Levenshtein distance between the first i characters of string A and the first j characters of string B can be calculated using the formula shown in Eq. 2:

$$lev_{a,b}(i,j) = \begin{cases} \max(i,j) \\ \min\begin{cases} lev_{a,b}(i-1,j)+1 \\ lev_{a,b}(i,j-1)+1 \\ lev_{a,b}(i-1,j-1)+1(a_i \neq b_j) \end{cases} \end{cases} \tag{2}$$

where l() is an indicator function that returns 1 when the ith character of string a is different from the jth character of string b, and 0 otherwise.

Here is the implementation of the Levenshtein distance algorithm:

Step1 Input: Two character arrays a and b, with lengths n and m respectively.
Step2 Output: Levenshtein distance $D_{[n,\ m]}$.
Step3 Define a two-dimensional array D with a size of (n + 1) × (m + 1) to store the minimum edit distance.
Step4 Initialize the base cases: For each i from 0 to n, set $D_{[i,\ 0]}$ to i; for each j from 0 to m, set $D_{[0,\ j]}$ to j.
Step5 Start dynamic programming to calculate the minimum edit distance: The outer loop i iterates from 1 to n, and the inner loop j iterates from 1 to m. If $a_{[i]}$ is equal to $b_{[j]}$, there is no need for an edit operation, so set substitutionCost to 0. If $a_{[i]}$ is not equal

to $b_{[j]}$, an edit operation is required, so set substitutionCost to *1*. Calculate $D_{[i, j]}$ by selecting the minimum edit distance among the deletion ($D_{[i-1, j]} + 1$), insertion ($D_{[i, j-1]} + 1$), and substitution ($D_{[i-1, j-1]}$ + substitutionCost) operations, and assign the result to $D_{[i, j]}$.
Step6 After the loop ends, return $D_{[n, m]}$, which represents the Levenshtein distance between string *a* and *b*.

4.3 Algorithm Combinations

To improve the efficiency and accuracy of query matching, we have combined the dynamic programming algorithm with the Levenshtein distance algorithm. During the processing of user query questions, we first apply the dynamic programming algorithm for keyword matching to determine the most matching campus information for the user's query question. Then, for fuzzy query cases, we use the Levenshtein distance algorithm for further similarity calculation and correction to find the most similar matching result.

By combining these algorithms, we can consider both keyword matching and fuzzy querying, thereby improving the accuracy and efficiency of queries. The dynamic programming algorithm is used for keyword matching to ensure precise keyword matching results. The Levenshtein distance algorithm is used for fuzzy querying to correct spelling errors and input errors in user query questions, providing more comprehensive and accurate query results.

This code is a backend code based on Node.js and Express. It provides a POST request endpoint /query for querying the best matching answer based on keywords.

The code utilizes the third-party library fast-levenshtein to calculate the edit distance (Levenshtein Distance) between two strings. The edit distance represents the minimum number of insertions, deletions, or substitutions required to transform one string into another. Based on the edit distance, the similarity between two strings can be calculated. Please refer to Fig. 7 for illustration.

The code retrieves the question parameter from the requested req.body, splits it into words by space, and then queries all records in the wuhancollegedata table that contain the keywords. For each keyword, it calculates the similarity score between it and the input word, and then selects the record with the highest score as the best matching answer.

It is important to note that the code uses a connection pool (pool.getConnection and connection.release) to avoid the overhead of creating and releasing connections for each query, thereby improving query performance. Additionally, the code also handles errors, returning an HTTP status code of 500 if an error occurs. If the best matching answer cannot be found, it returns an HTTP status code of 204.

Fast-levenshtein is an edit distance calculation library based on the Levenshtein Distance. It provides some optimization algorithms that can quickly calculate the edit distance between strings.

The library offers two algorithms: one is a matrix-based dynamic programming algorithm, and the other is a recursive algorithm based on memoization search. Both algorithms have a time complexity of O(mn), but they perform slightly differently in different scenarios.

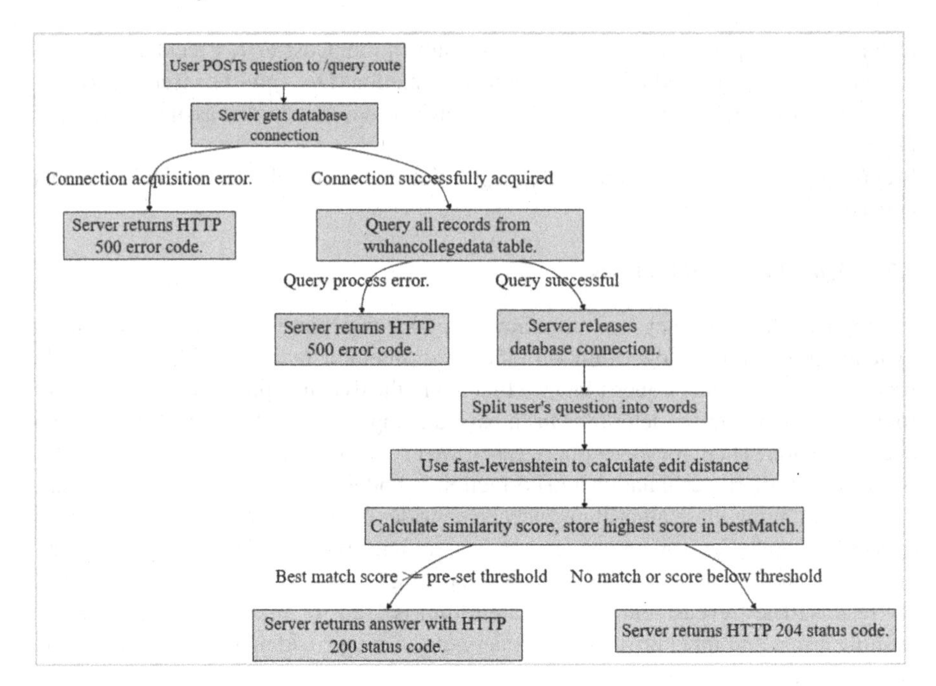

Fig. 7. Code flowchart

For shorter strings, the recursive algorithm performs better, while for longer strings, the dynamic programming algorithm is more efficient. Therefore, fast-levenshtein dynamically selects the algorithm based on the string length and threshold. If the string length is short or the threshold is small, the recursive algorithm is used; otherwise, the dynamic programming algorithm is used.

Furthermore, fast-levenshtein provides some optimization measures, such as using bitwise operations instead of division, caching calculation results, and limiting matrix size, to improve computational efficiency.

4.4 Process of Algorithm and Interfaces

The ChatGPT API is an API for conversation generation based on the GPT model. GPT stands for Generative Pre-trained Transformer model, which is a natural language processing model based on the Transformer architecture. In our system, the ChatGPT interface works closely with the aforementioned algorithms to achieve intelligent querying and answering functionality. Please refer to Fig. 8 for illustration.

The calling process is as follows:

The user inputs a query question and submits it to the system.

The system first applies the dynamic programming algorithm for keyword matching, based on the user's query question and the campus information in the database, to determine the best matching result.

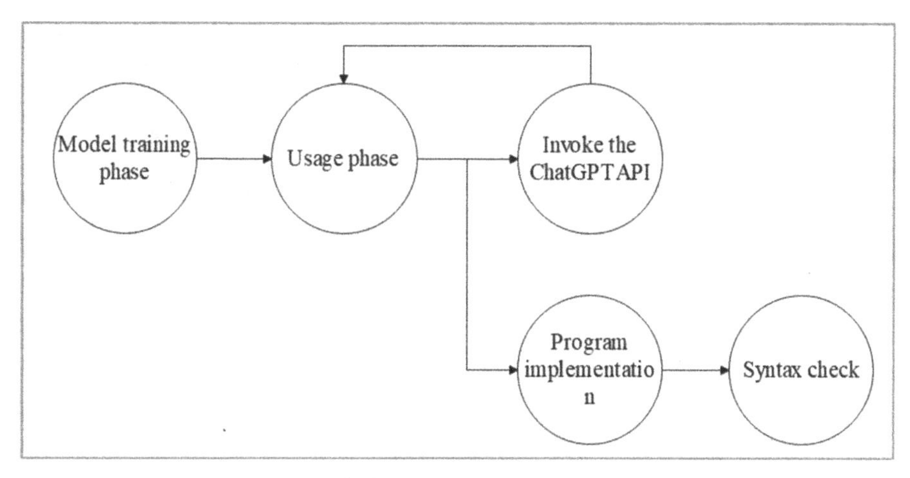

Fig.8. Algorithm flowchart

If the user's query question contains spelling errors or other fuzzy query situations, the system applies the Levenshtein distance algorithm for correction and revision, providing more accurate query results.

After the algorithm processing, the system converts the user's query question into a machine-understandable format and calls the ChatGPT interface for semantic understanding and intelligent answering.

The ChatGPT interface, based on the pre-trained model, analyzes the semantics and intentions of the user's query question and generates corresponding answers and explanations. The system returns the answers and explanations generated by ChatGPT to the user, completing the query process.

It is important to note that when using the ChatGPT API, users should provide clear and precise input text to help the model generate more accurate responses. Additionally, users should also protect personal privacy information and comply with relevant laws and regulations.

getChatGPTResponse(message)
This function is used to send a request to the ChatGPT API to get a reply to the user's input text (message). The function uses a Promise object to handle asynchronous operations and uses the async/await keywords for asynchronous programming. Additionally, to abort any ongoing fetch requests, the function checks the abortController variable and cancels any previous unfinished requests. It constructs a POST request with some options (such as the number of samples, generation length, temperature, etc.) and returns the answer generated by the API.

checkGrammar(answer)
This function is used to check if there are any grammar issues in the information (answer) generated by getChatGPTResponse, and performs some error correction. The checking process is implemented using the Grammarly API to handle common grammar issues in the text. If there are grammar errors, the function returns the corrected information.

Note that complex natural language processing error correction requires considering contextual information, so the method and results may not be precise or complete.

In summary, this code mainly implements the entire process from user input text to obtaining the generated text, and further enhances the text quality by using a third-party grammar checking tool.

Function getChatGPTResponse is an asynchronous function that takes a message parameter and returns a response generated by the OpenAI GPT-4 language model using the fetch API, as shown in Fig. 9.

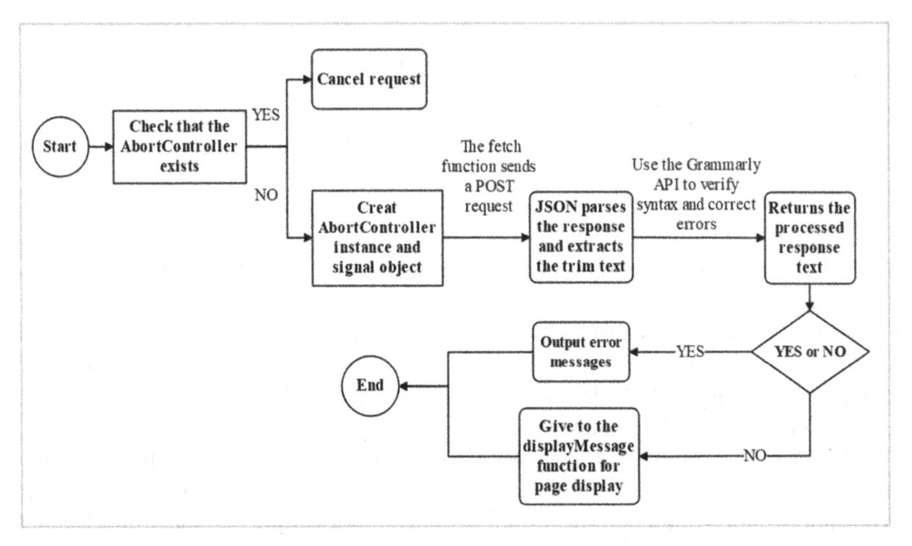

Fig. 9. Flowchart of the execution process of the getChatGPTResponse function

Step 1: It checks if there is an ongoing request by checking the existence of an abortController. If it exists, the request is canceled. Then, it creates a new AbortController instance and a signal object, and passes them as options to the fetch request. This allows for canceling the request when necessary.

Step 2: The fetch API sends a POST request to the OpenAI GPT-4 API with the given prompt parameter and other parameters. Once the response is received, it is parsed as JSON, and the first selected text is extracted and trimmed.

Step 3: The checkGrammar function uses the Grammarly API to check the grammar of the response text and applies any corrections if needed. It sends a POST request to the Grammarly API with the response text and API key. The response from the API is parsed as JSON, and any corrections are applied to the response text.

Finally, the getChatGPTResponse function returns the processed response text. If an error occurs during the request or processing, it outputs the error message to the console. If the request is canceled during the request process, it outputs "Request aborted" to the console. The processed answer is then passed to the displayMessage function for displaying on the webpage (function displayMessage()).

4.5 Advantages of the Algorithm

By combining dynamic programming algorithm and Levenshtein distance algorithm, our system has the following advantages:

Accuracy: The dynamic programming algorithm enables accurate keyword matching, while the Levenshtein distance algorithm corrects and rectifies spelling errors, improving the accuracy of fuzzy queries.

Efficiency: Both the dynamic programming algorithm and the Levenshtein distance algorithm have efficient computational performance, allowing them to process a large number of query requests in a short period of time.

Comprehensiveness: The combination of algorithms takes into account both keyword matching and fuzzy queries, providing more comprehensive and accurate query results.

Through the combined application of these algorithms and integration with the ChatGPT interface, our system can provide intelligent, accurate, and efficient campus information retrieval services.

4.6 Results

Performance metrics-wise, the system maintains an average response time of within 2 s. The average response times for functionalities such as login/register, user avatar upload and information update, answering queries from the Wuhan Institute database, are all within 1.5 s. The average response time for ChatGPT's answers is 3 s. Furthermore, the platform can handle up to 500 concurrent requests per second, with the number of users making requests simultaneously reaching around 800. The query rate can reach approximately 200 queries per second. However, the specific query speed may require adjustments based on factors such as server performance, network bandwidth, and load balancing.

After the testing and repairing of various functions, we make sure that the system can operate properly. Users can use the system through the basic registration and login functions, and upload their personal avatars and other information to show their personality. Users can talk to our little robot to quickly get campus information. Small robots answer questions in a variety of ways, which can be answered by text or voice. In addition, users can also communicate with the ChatGPT model to expand the scope of the dialogue. Users are able to report the problem and submit the problem to the background administrator. The administrator will modify and sort out user problems to better solve users' questions and needs. Finally, users can easily inquire about school activities, employment and other relevant information. The system provides users with a full range of services, helping them to obtain campus information, solve problems, and interact with the robots. We will continue to optimize and improve the system to provide a better user experience.

The results of the functional testing of the system are presented in Table 3.

Table 3. Test Results

Functions	Test description	Defect situations	Fix the situation	Test results
User avatar upload and modification	Upload or modify avatar	User avatar images keep accumulating upon upload	Automatically delete previous avatars	Successfully deleted the historical avatar
Real-time updating of conversations	Continuation of conversation	After sending a question, the system remains in a waiting state	Cancel the previous request and focus on the current request	Successfully updated the conversation in real-time
Message display and formatting	User interface beautification	Issues with the formatting and layout of user avatars and text messages	Keep the user avatar and username fixed above the message box	Successfully beautified the interface
ChatGPT API integration	Network requirements for API calls	High network requirements for users	Move the API calls to the server-side	Successfully resolved high network requirements
Post-processing of ChatGPT responses	Stability of responses	Randomness and instability in the system's responses to API calls	Set key attributes and perform post-processing	Successfully optimized the stability of responses
Integration with the Wuhan College information database	Proper display of information retrieved from the database	Images are not displayed properly	Modify the logic for storing images and return the image path	Successfully called the function, and now text and images can be displayed normally
keyword matching algorithm	Keyword matching, fine-tuning the most suitable threshold	None	Adjust the threshold to 0.76	Successfully matched user keyword queries

5 Conclusion

5.1 Summary of the Paper

This paper designs and develops an enhanced campus information query system based on the ChatGPT interface and local content database. By researching the advantages and applications of ChatGPT in intelligent dialogue and natural language processing, we propose a system design and algorithm combination scheme. In the system design part, we introduce the architecture design, module hierarchy, interface design, and database design of the system. In the algorithm combination part, we apply dynamic programming algorithm and Levenshtein distance algorithm to achieve keyword matching and fuzzy querying. Through the implementation of the system, we are able to provide intelligent, accurate, and efficient campus information query services.

5.2 Advantages and Prospects

Through the research and implementation of this paper, our campus information retrieval system has the following advantages:

Intelligence: By utilizing the ChatGPT interface, the system is capable of intelligent conversations and answers, providing personalized query services and recommendations.

Efficiency: With the efficient computing performance of dynamic programming algorithms and Levenshtein distance algorithm, the system can handle a large number of query requests in a short time.

However, there are still aspects of our system that can be further improved. Future research can focus on the following directions:

Algorithm optimization: Further optimize the dynamic programming algorithm and Levenshtein distance algorithm to improve query matching efficiency and accuracy.

Database expansion: Increase the campus information resources in the database, covering more fields and disciplines to provide more comprehensive and rich query results.

Cross-platform applications: Extend the system to more platforms and devices, such as mobile applications, smart speakers, etc., to provide more convenient and seamless campus information retrieval services.

References

1. Gao, X.: Design and implementation of parent-child system based on uni-app+express. Comput. Inf. Technol. **31**(02), 49–52+58 (2023). https://doi.org/10.19414/j.cnki.1005-1228.2023.02.012
2. Feng, Z., Zhang, D., Rao, G.: From turing test to ChatGPT: milestones and insights in human-machine conversation. Lang. Strateg. Res. **8**(02), 20–24 (2023)
3. Kim, B.: ChatGPT and generative AI tools for learning and research. Comput. Libr. **43**(6) (2023)
4. Liu, K.: Exploring the application potential of the large language model in sociological research: a case study of ChatGPT. Soc. Sci. Humanit. Sustain. Res.**4**(3) (2023)
5. Mohd, J., Abid, H., Pratap, R.S., et al.: Unlocking the opportunities through ChatGPT tool towards ameliorating the education system. BenchCouncil Trans. Benchmarks Stan. Evaluations, **3**(2), 100115 (2023)
6. Li, K., Wang, D., Xing, C., et al.: Design and Implementation of intelligent financial customer service system for universities. In: Network Application Branch of China Computer Users Association. Proceedings of the 26th Annual Conference on Network New Technologies and Applications of China Computer Users Association [Publisher unknown], pp. 151–153 (2022). https://doi.org/10.26914/c.cnkihy.2022.049267

7. Yan, S., Fu, L., Xing, Y., et al.: Design and implementation of campus recruitment intelligent customer service based on seq2seq. Audio Eng. **46**(08), 72–74+82 (2022). https://doi.org/10.16311/j.audioe.2022.08.021
8. Li, T., Cui, Q., Li, X.: Fuzzy matching algorithm based on word frequency weighting and cosine similarity. Enterp. Sci. Technol. Dev. **493**(11), 49–51 (2022)
9. Qin, M.: Design and implementation of data visualization system based on B/S mode. [Dissertation]. Beijing Univ. Posts Telecommun. (2020). https://doi.org/10.26969/d.cnki.gbydu.2020.001179
10. David, C.: GPU acceleration of levenshtein distance computation between long strings. Parallel Comput. **116**, 103019 (2023)

Prediction of TrkB Complex and Antidepressant Targets Leveraging Big Data

Xufu Xiang(✉), Chungen Qian, Xin Liu, and Fuzhen Xia

The Key Laboratory for Biomedical Photonics of MOE at Wuhan National Laboratory for Optoelectronics - Hubei Bioinformatics and Molecular Imaging Key Laboratory, Systems Biology Theme, Department of Biomedical Engineering, College of Life Science and Technology, Huazhong University of Science and Technology, Wuhan 430074, China
xiang_xufu@hust.edu.cn

Abstract. Since Major Depressive Disorder (MDD) represents a neurological pathology caused by inter-synaptic messaging errors, membrane receptors, the source of signal cascades, constitute appealing drugs targets. G protein coupled receptors (GPCRs) and ion channel receptors chelated antidepressants (ADs) high-resolution architectures were reported to realize receptors physical mechanism and design prototype compounds with minimal side effects. Tyrosine kinase receptor 2 (TrkB), a receptor that directly modulates synaptic plasticity, has a finite three-dimensional chart due to its high molecular mass and intrinsically disordered regions (IDRs). Leveraging breakthroughs in deep learning, the meticulous architecture of TrkB was projected employing Alphafold 2 (AF2). Furthermore, the Alphafold Multimer algorithm (AF-M) models the coupling of intra- and extra-membrane topologies to chaperones: mBDNF, SHP2, Etc. Conjugating firmly dimeric transmembrane helix with novel compounds like 2R,6R-hydroxynorketamine (2R,6R-HNK) expands scopes of drug screening to encompass all coding sequences throughout genomes. The operational implementation of TrkB kinase-SHP2, PLCγ1, and SHC1 ensembles has paved the path for machine learning in which it can forecast structural transitions in the self-assembly and self-dissociation of molecules during trillions of cellular mechanisms. In silicon, the cornerstone of the alteration will be big data and artificial intelligence (AI), empowering signal networks to operate at the atomic level and picosecond timescales.

Keywords: Major Depressive Disorder (MDD) · Antidepressants (ADs) · Tyrosine kinase receptor 2 (TrkB) · Alphafold Multimer algorithm (AF-M) · Big Data · Artificial Intelligence (AI)

1 Introduction

MDD, a systemic psychiatric disorder, may have a wide range of implicit etiologies: synaptic misconnection, metabolic abnormalities and immune inflammation. The lifetime prevalence of MDD surpasses 20 % in the worldwide population, and the unavailability of specific medications renders one-third of patients unresponsive to treatment

S. Zhang et al. (Eds.): BigData 2023, LNCS 14203, pp. 149–165, 2023.
https://doi.org/10.1007/978-3-031-44725-9_11

[1]. As the initial spot of the signal cascade, transmembrane receptors are indispensable in gaining and transferring signals across a thousand trillion interconnected synapses and as the objective of fifty percent of prescription medicine [2, 3]. Several ADs have a direct link to the 5-hydroxytryptamine receprots (5-HTRs) of GPCRs and the ion channel-type glutamate receptors NMDAR and AMPAR for the onset of action [4]. Subtypes of 5-HTRs binding maps with ADs have been constructed then using virtual pharmacological screenings of millions of compounds to catch novel non-hallucinogenic antidepressants [5–9]. The results of conjugating NMDAR with S-ketamine reveal that S-ketamine takes effect more rapidly than conventional ADs by inhibiting Ca^{2+} influx to intracellular [10]. Previously, the predominant hypothesis claimed that ADs indirectly positively regulate TrkB via NMDAR to trigger the synaptic plasticity mechanism [11]. However, in a latest report, robust R-ketamine release therapeutic advantages by directly binding to TrkB [12]. This casts doubt on the broadly held 5-HT and NMDAR hypotheses and propels TrkB to spring up as a momentous priority receptor in building neat antidepressants.

Determining protein structure is essential for pharmaceutical research, while there is a deficit of architectural insights into TrkB despite the profusion of foundational research results. TrkB modulates postsynaptic protein expression and synaptogenesis through the MAPK, PIK3/mTOR, and PLC pathways, intimately associated with a multitude of psychiatric conditions, including depression, Parkinson's disease, and schizophrenia. Like many other RTKs, due to their variable transmembrane topology and molecular mass outweighing the upper limit of conventional resolution approaches, only fifty percent of TrkB sequences have solved. Several RTKs, including EGFR, INSR, and ALK, have accurate measurements of the extracellular segment thanks to recent methodological enhancements in Cryo-electron microscopy (Cryo-EM). However, the membrane-spanning and cytoplasmic sections of all ligand-binding multimers were presented in low resolution due to IDRs. The dearth of architectural information has sparked debate on a variety of hot-button issues, along with the following: Why the extracellular TrkB segment reacts to distinct NGF-family members having diverse biological responses? Does TrkB's transmembrane helix bridging angle alter the on/off state and strength of its subsequent enzymatic activity, and is it a feasible curative landing point?How does TrkB's intracellular kinase element engage over 140 chaperones to assemble signal gatherings and automatically disentangle post-phosphorylation?Only a few layouts of RTKs bound to key partners were fixed since resolution entails a rigorous selection of docking sequences, kinase phosphorylation phase manipulation, and crystallographic strategies. AF2, a deep learning-based artificial intelligence algorithm, predicted and released the structures of all protein sequences, making them easily accessible before resolution. Previously, the GPCRs and ATP-binding cassette receptors were correctly identified as the experimental structures adopting AF2. RTKs predicting topologies are rarely utilized because of their trisecting topologies and characterization of dimer activation. As a result of AF2 advancements, biologists can construct biochemically active aggregates utilizing the protein complex prediction algorithm AF-M. Even if TrkB is unresolved, AF-M can build its signal assemblies, and clear out if the TrkB dimeric transmembrane helix's architecture does bind to ADs.

In the first step, utilizing AF2, the full-length monomer of TrkB was built with atomic-level precision. Due to IDR, the full-length dimeric structure of TrkB was not topology-compliant. To tackle this problem, the TrkB sequence was divided into extracellular, transmembrane, and intracellular segments. TrkB extracellular portion relating to mBDNF has predicted, and it is consistent with the resolved structure possessing the similar intrinsic and specific ligand-binding regions. Creating the model of TrkC linking NTF3 indicates that AF-M rapidly pair between homologous receptors and ligands. The structural information help design microproteins with regulatory activity for a specific pair. In physiological membrane environment, the transmembrane helix dimers maintained stable, and binding pocket for the novel ADs: IHCH-7086 and (2R, 6R)-HNK was localized in the helix crossing point. This discovery expands the list of pharmacological targets from GPCRs and ion channel-type receptors to include RTKs. AI detected the essential posture of the assemblies during phosphorylation using related signal substrates like PLCγ1, SHP2, and SHC1. In conclusion, the TrkB's snapshots during the signal cascade have been effectively reproduced using AF-M. Networks of physiological or pathological signal pathways comprising spatiotemporal information will be built in computers with artificial intelligence. In the next generation, AI will provide crucial points of the unresolved protein architectures associated with pathologies, broadening the scope of structure-based drug discovery to all coding sequences.

2 Methods

2.1 AF2 Predicts the Full-Length Structure of TrkB

To predict the full-length structure of TrkB, the first one was obtained directly from the AF Protein Structure Database (Last updated in Alphafold database, version 2022–06–01, created with the AF Monomer v2.0 pipeline). The second one is based on AF Monomer V2.2, manually inputting the full-length sequence of TrkB obtained from uniport. The third one is based on the monomer_casp14 program, which improves the average GDT of Monomer by about 0.1. AF2 was downloaded from github and run locally as described (https://github.com/deepmind/AF), using default parameters and the database version used: pdb _mmcif and uniport are 2022–08–03, the rest of the database is the default database.

2.2 AF-Multimer Predicts the Structure of Protein Complexes

Human full-length sequences of TrkB, BDNF, SHP2, PLCγ1, and SHC were obtained from uniport, and sequences were selected for combination to construct multimers as needed, and the sequence combinations used are shown in the Supplementary Material. Run the AF-multimer program and set '--model_preset = multimer' to output the PDB file of the top 5 predicted complex structures sorted by PLDDT. All raw structures not shown are shown in the supplementary material.

2.3 Conformational Optimization of Drug Small Molecule Drawing

The (2R, 6R)-HNK, IHCH-7086 structures were drawn using ChemDraw, calling the Chem.AllChem module of RDKi (http://rdkit.org) using the Embed Molecule function using Experimental-Torsion Basic Knowledge Distance Geometry (ETKDG) algorithm to generate 3D conformations based on the modified distance geometry algorithm, optimize and calculate the energy using MMFF94 force field, and finally select the lowest energy conformation as the initial conformation for docking.

2.4 Protein-Drug Molecule Docking

The highest confidence 427–459 double alpha helix structure obtained by AF-M was taken, and Smina was selected as the docking software for molecular docking with (2R, 6R)-HNK and IHCH-7086. The positions of the active centers were as follows: X-center = −7.729, Y-center = 2.684, and Z-center = −6.868, where the approximation (exhaustiveness) of docking was 80, the box size of docking was 40 Å, and 80 Å conformations were generated each time, and the optimal conformations were selected for molecular dynamics simulations.

2.5 Analysis of Protein-Ligand Interactions

Upload the pdb file of protein complex to PILP for interaction analysis, set the A chain as the main chain, obtain the salt bond, hydrophobic bond, ππ bond, etc. generated with the interacting ligands, and visualize the output pse file with pymol.

2.6 Sequences Alignment

Clustal Omega is used to perform multiple sequence alignment, input the sequence information obtained from uniport, and set the output form as ClustalW with character counts. Visualizing the comparison results using Jalview. The higher the similarity, the darker the color of the residues.

2.7 Molecular Dynamics Simulation

The conformation of ranked 1 transmembrane helical dimer coupled with (2R, 6R)-HNK predicted by AF-M was added with 20% CHOL + 80% DOPC as the membrane environment. Gromacs2019.6 was chosen as the kinetic simulation software and amber14sb as the protein force field. Small molecules were used to produce topology files based on GAFF2 (Generation Amber Force Field) force field. The TIP3P water model was used to add TIP3P water model to the complex system to create a water box and add sodium ions to equilibrate the system. Under elastic simulation by Verlet and cg algorithms respectively, PME deals with electrostatic interactions and energy minimization using steepest descent method for maximum number of steps (50,000 steps). The Coulomb force cutoff distance and van der Waals radius cutoff distance are both 1.4 nm, and finally the system is equilibrated using the regular system (NVT) and isothermal isobaric system (NPT), and then the MDS simulations are performed for 100 ns at room temperature and pressure. In the MDS simulations, the hydrogen bonds are constrained by the LINCS algorithm with an integration step of 2 fs. The Particle-mesh Ewald (PME) method is calculated with a cutoff value of 1.2 nm, and the non-bond interaction cutoff value is set to 10 Å. The simulation temperature is controlled by the V-rescale temperature coupling method at 300 K, and the pressure is controlled by the Berendsen method at 1 Å. NVT and NPT equilibrium simulations were performed at 300 K for 30 ps, and finally, the finished MDS simulations were performed for 50 ns. The root mean square deviation (RMSD) was used to observe the local site variation during the simulation (the cut-off point was set to 0.2). The radius of gyration (Rg, radius of gyration) is used to evaluate the tightness of the structure of the system. The root mean square function (RMSF) is used to observe the local site metastability of the system during the simulation, and the solvent accessible surface area (SASA) is used to observe the size of the solvent accessible surface area of the complex during the simulation.

2.8 Binding Free Energy Calculation for Proteins and Small Molecules

The MDS trajectory is used to calculate the binding free energy by the following equation:

$$\begin{aligned}\Delta G_{bind} &= \Delta \mathrm{G}_{\mathrm{complex}} - \left(\Delta \mathrm{G}_{\mathrm{receptor}} + \Delta \mathrm{G}_{\mathrm{ligand}}\right)\\ &= \Delta \mathrm{E}_{\mathrm{internal}} + \Delta \mathrm{E}_{\mathrm{VDW}} + \Delta \mathrm{E}_{\mathrm{elec}} + \Delta \mathrm{G}_{\mathrm{GB}} + \Delta \mathrm{G}_{\mathrm{SA}}\end{aligned}$$

In the above equations, $\Delta E_{internal}$ internal represents internal energy, ΔE_{VDW} represents van der Waals interaction and ΔE_{elec} represents electrostatic interaction. The internal energy includes the bond energy (Ebond), angular energy (Eangle), and torsion energy (Etorsion); ΔG_{GB} and ΔG_{GA} are collectively referred to as the solvation free energy. Among them, GGB is the polar solvation free energy and GSA is the nonpolar solvation free energy. For ΔG_{GB}, the GB model is used for calculation (igb = 8). The nonpolar solvation free energy (ΔGSA) is calculated based on the product of surface tension (γ) and solvent accessible surface area (SA), GSA = 0.0072 × SASA. The entropy change is neglected in this study due to the high computational resources and low precision. This study is neglected. This algorithm is implemented by Gmx_MMPBSA.

3 Result

3.1 AF2 Predict the Full-Length and Dimeric Structures of TrkB

- Predicting the full-length structure of TrkB monomers by three methods using AF2

The fixed structures predicted by the three modalities were similar, TrkB consists of five major structures, Domain 1: two CR clusters sandwiching three LRRs, Domain 2: Ig C1, Domain 3: Ig C2, Domain 4: transmembrane α-helix, and Domain 5: kinase domain (Fig. 1A). However, the main difference between the predicted results could be attributed to the disordered sequences on both sides of the transmembrane helix and at the N and C-terminals. The PLDDT values of these sequences are shown to be low and as orange noodles by AF2. The disordered sequences are freely distributed, resulting in the predicted extracellular, transmembrane, and intracellular relative positions of TrkB not conforming to their spatial distribution and separated from each other at the plasma membrane.

- AF-M predicts the full-length dimer structure of TrkB without ligand/binding ligand and predicts the spatial distribution of the structure irrationally

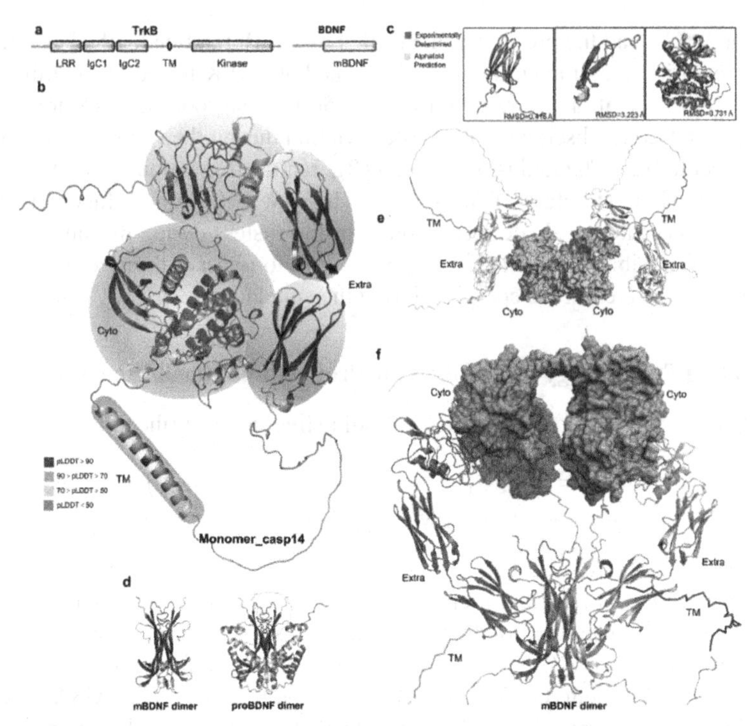

Fig. 1. TrkB monomer, dimer structure predicted by AF2. (A) Schematic diagram of the structural domains of TrkB and BDNF. (B) The full-length structure of TrkB predicted by casp14, with different interval confidence levels displayed in the structure in the corresponding colors. Fixed structures are circled and colored in line with (A). (C) Structural alignment between the AF predicted structure and the experimentally resolved structure, in order: IgC2 [PDB:1WWB], juxtamembrane IDR [PDB:2MFQ], kinase domain [PDB:4ASZ], the RMSD between the predicted structure and the experimental structure is marked in the lower right corner, respectively. (D) mBDNF and proBDNF homodimeric structures predicted by AF-M. (E) The ligand-free dimeric structure of TrkB predicted by AF-M does not possess a reasonable membrane topology. (F) AF-M predicted structure of the activation dimer of TrkB-binding mBDNF. (Color figure online)

AF-Multimer, used for protein complex prediction, has been released recently and has demonstrated excellent performance in constructing dimers and multimers. It has been used for dimer construction of TrkB to understand the usability of AF2 for RTKs, which have dynamic conformations and a wide range of binding partners and complex binding forms. The structure of TrkB has not yet been resolved at high resolution under Cryo-EM, and AF2 does not allow direct access to experimental dimer structures. Nevertheless, the extracellular dimer and transmembrane dimer of the TrkB homologous protein TrkA have been resolved to use it as a reference for the predicted structure of TrkB to evaluate the availability of AF-M predictions.

The homodimers of the two forms of BDNF were constructed independently (Fig. 1D). mBDNF dimer is highly similar in structure to the already resolved mBDNF/NTF-4 heterodimer 55. The main body of mBDNF consists of three pairs of β-strands and four linked β-hairpin loops, and the dimer is centrosymmetric with the long axes of two pairs of long β-strands contact (Fig. 1D). In contrast, the structure of the proBDNF dimer has not yet been resolved, and the predicted structure has four incomplete helices at the N-terminal of each monomer. The presence of these helices caused the lower end of the β-strands to be pulled to both sides, which disrupted the intrinsic binding interface of the NGF-β family at the IgC2 of the Trk family. This alteration revealed the reason for the low affinity of proBDNF to TrkB and high affinity to p75, providing a structural explanation for the opposite synaptic effect of proBDNF and mBDNF. Consequently, AF predicted mBDNF homodimers, demonstrating that it is highly usable in the prediction of dimeric ligands and can rapidly predict all the putative intrafamily paired dimer conformations when homologous family dimeric structures are available.

Based on the successful construction of mBDNF dimer, we further predicted the conformation of TrkB full-length dimeric activation state, TrkB-mBDNF-mBDNF-TrkB. The major dimerization interface of all predicted results was the extracellular segment, and the kinase dimerization structure in the ligand-free dimerization structure was separated, indicating centrosymmetric but no contact. In ranked 1 as an example: mBDNF bridges the bipartite structure of the extracellular segment, while the kinase segment is incorrectly placed at the top of the N-terminal of the extracellular segment and does not contact each other (Fig. 1F). The transmembrane helix is recognized as a disordered structure, indicating that the addition of mBDNF does not rescue the irrational spatial distribution of TrkB dimeric conformation predicted by AF-M.

In conclusion, we found that AF-M does not provide a reasonable full-length dimeric structure of TrkB with or without binding ligands. This phenomenon is attributed to the fact that the resolution results of RTKs Cryo-EM could not provide a sufficient density for the transmembrane and intracellular segments due to the lack of rigid connections between the extracellular segment and transmembrane helix, and only the dimeric extracellular segment was resolved at high resolution and uploaded to the database. Despite the inability to give a full-length dimeric structure, AF-M provides a dimeric conformation of the kinase segment and the extracellular segment of the bound ligand, derived from the mimicry of the local structures of RTKs obtained by NMR and X-ray over the past 30 years. This indicates that AF-M has the ability to predict the fixed structure of RTKs undergoing dimerization. Hence, we split the sequence of TrkB into extracellular/transmembrane helix/intracellular segments for dimerization prediction based on previous studies of EGFR and predicted their complex conformation upon binding ligands, signal components, or drugs.

3.2 AF-M Accurately Predicts the Ligand-Binding Structure of the Extracellular Segment of the TrkB Dimer

- Prediction of potential TrkB extracellular ligand-free binding dimer structure with inter-monomeric binding via β-turn

The predicted TrkB extracellular ligand-free dimer was obtained by AF-M. The docking of ranked 1–ranked 3 occurs at Domain 3, while the difference between the structures is that the two monomers Domain 3 are increasingly spaced apart with decreasing confidence until no contact is made; the docking of ranked 4–5 occurs at Domain 1 and IDR. These findings differed from the previously experimentally obtained dimeric structure of Trk family Domain 3, wherein the crystal structure is based on the two monomers with overlapping N-terminus and C-terminus, which results in the loss of βA at the N-terminus of Domain 3. Such a structure is considered erroneous because Domain 2 exists on Domain 3 N-terminal, and the N-terminal sequence serves as a bridge between Domain 2 with 3. Despite the incorrectly constructed Domain 3 dimer in the database, AF-M can determine and correct its implausibility. In ranked 1, βA was successfully identified and linked to Domain 2, and the monomers contacted each other through the loop between βA and βB (ABL), i.e., the interaction force of S297-W301 (Fig. 2A and 2B). A total of five hydrogen bonds and two hydrophobic bonds were formed between the residues, and a π-π stacking was formed between H299 and W301. These residues formed a negatively charged finger-like protrusion and a concavity, and the two monomers formed a chimeric structure with each other, which was consistent with the subsequent prediction of a specific ligand-binding region (Fig. 2C). Furthermore, M379 and G380 near the membrane were also identified as bind residues in the disordered region. AF-M offers the possibility of an unresolved TrkB pre-dimer, i.e., ABLs that contact Domain 3 specifically with ligands forming chimeras with each other, and the ligand-binding surface is buried until the ligand is inserted and opened to stimulate the downstream signals.

- Predicting the ligand-bound dimeric structure of TrkB extracellularly with specific binding region and intrinsic binding region between Domain 3 and mBDNF

Next, AF-M was used to predict the conformation of the extracellular segment of TrkB bound to mBDNF. The five results showed a high degree of agreement, with differences arising from the free distribution of the disordered region. The complex structure of mBDNF retains its centrosymmetric homodimeric conformation and binds to the Domain3 of TrkB. Conversely, the extracellular segment of TrkB has a crabpincer shape with two monomers each monomer is attached to a single chain, alternating between the front and back of mBDNF. This finding is similar to the results of the TrkA extracellular segment. The most different secondary structure is Domain 1, wherein β1–β3 starting from the N terminus is shorter in the crystalline structure than in the predicted structure, and most of the articulated sequences have a loop-like morphology (Fig S5B). These results predicted a TrkB Domain 1 superhelical topology compared to the resolved TrkA.

In conclusion, in the structure prediction of extracellular segment-binding ligands, AF-M is unable to predict the effect of PTMs on the structure, resulting in a relative angle between domains deviating from the true structure. However, the predicted structures are highly accurate for the ligand-bound receptor structural domains, constructed with reference to the experimental structures between ligand-same family receptor members of the same family in the database. Previous studies have shown that several RTKs can form heterodimeric pairings with homologous receptors. Also, ligand family members can heterodimerize, with dozens of possible pairings between the Trk family and the NGF-β family alone. AF-M facilitates the assembly of all the ligand-receptor binding structures using only the sequences, thereby circumventing the limitation of the resolved structures. Thus, we can retrieve highly matched design structure microproteins and small-molecule drugs using the three-dimensional structural information of proteins, thereby minimizing the side effects of the traditional antibody binding to the same family of receptors and regulating the activation and inactivation of receptor kinases at the atomic level.

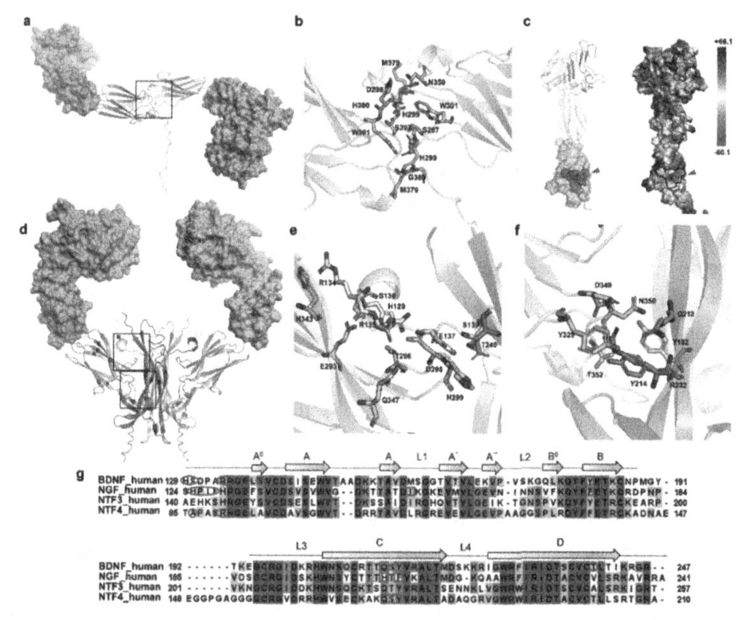

Fig. 2. Structure prediction of pre-activated and activated dimer of TrkB extracellular segment. (A) AF-M prediction of the potential TrkB extracellular segment pre-activated dimer structureb. (B) Detailed graphical representation of the contact residues of the pre-activated TrkB pre-dimer. (C) TrkB pre-dimer contact residues form finger-like protrusion and concavity, marked in purple (left). Circled in black stroke on the surface potential energy map (right). (D) AF-M prediction of the extracellular activation state dimer of TrkB that binds mBDNF. The two interact regions are framed in black. (E) Residue details of the TrkB-binding mBDNF-specific binding region are shown. (F) Detail map of residues in the intrinsic binding region of TrkB-binding mBDNF is shown. (G) Sequence alignment of NGF-β family, where residues in the specific binding region are circled and residues in the intrinsic binding region are boxed. Information of BDNF is from the predicted structure of TrkB-mBDNF. Information of NGF NTF4 is from the resolved structure of TrkA-NGF and TrkB-NTF4.

3.3 AF-M Predicts the Antidepressant Binding Pocket at the Crossover of TrkB Transmembranesegment Dimer

- The predicted TrkB transmembrane segment dimer structure is significantly different from the resolved structure of TrkA.

The transmembrane helix dimer of TrkB was successfully constructed by AF-M (Fig S6A). The transmembrane helix dimerization structure of TrkA has been resolved previously. The predicted transmembrane helix crossover pattern of TrkB is significantly different from that of TrkA, with TrkA monomers further apart from each other and the overall structure similar to X-type. On the other hand, the structure of TrkB tends to be parallel, and the helix of TrkB is longer than that of TrkA, with the crossover site ASVVG located at the center of the helix, while the crossover site of TrkA is near the N-terminal. Sequence comparison revealed that the crossover site SXAVG of TrkA and TrkC was shifted up by 7 residues compared to TrkB, indicating that the crossover site of TrkB is highly specific in the Trk family (Fig. 3D). The sequence comparison of TrkB between different species revealed that the transmembrane helix is conserved in the family. While the transmembrane helix is a common drug binding site for GPCRs and ion channel-type receptors, the complex formation structures of various antidepressants with the 5HT family and the glutamate receptor family have been resolved. Previous studies on EGFR have shown that the transmembrane helix transmits signals from the extracellular segment to the intracellular segment by changing the rotation angle and docking site. This alteration implies that the transmembrane helix of TrkB has the potential to modulate the activation strength of the intracellular kinases and becomes a target structure for antidepressants.

- Molecular dynamics simulations reveal a potential antidepressant binding pocket at TrkB docking.

The transmembrane helix structures predicted by AF-M were derived from the imitation of similar structures in the database. The AF-predicted transmembrane helix structures have been demonstrated to be highly accurate based on the comparison between the GPCR and ABC proteins with the resolved structures. The five TrkB transmembrane helix dimerization structures predicted by AF-M differed only in relative angles, with consistent crossover sites, indicating that TrkB dimers are formed in cholesterol-rich lipid rafts. To demonstrate the stability of the predicted structures in the membrane environment, the predicted ranked 1 dimer helix was placed in a 20% CHOL + 80% DOPC environmental box, and molecular dynamic simulations were performed for 100 ns, and the dimer structures were found to be cross-stable until the end of simulation.

In order to obtain the interaction of the drug with the transmembrane helix in the physiological membrane environment, the (2R, 6R)-HNK docked dimeric transmembrane helix structure was subjected to MDS in 20% CHOL + 80% DOPC environment (Fig. 3C). We found that the transmembrane helix showed a tendency to converge and reach a steady state in the second half of the simulation. The RMSD values fluctuate from 0–50 ns, which is caused by the initial instability of the docked acquired conformation inside the box, which declines rapidly at 50 ns and enters the steady state at 60 ns. The fluctuation of Rg corresponds to the ripple correspondence in RMSD, wherein the fluctuation rises in the first period, has a small peak at 10 ns, starts to reach a second peak at 50 ns and finally enters the stable zone at 75–100 ns, indicating that the system is shifting from the unstable to the stable state at this moment. The SASA of the protein gradually decreases in 0–100 ns, indicating a favorable binding and gradual protein tightening. HBNUM showed 0–2 connections of hydrogen bonding during the simulation.

Postural fingerprinting of the stable conformation revealed that the interaction of (2R, 6R)-HNK with the dimeric transmembrane helix originates from hydrogen bonding and the surrounding hydrophobic amino acids, whereby (2R, 6R)-HNK produced hydrogen

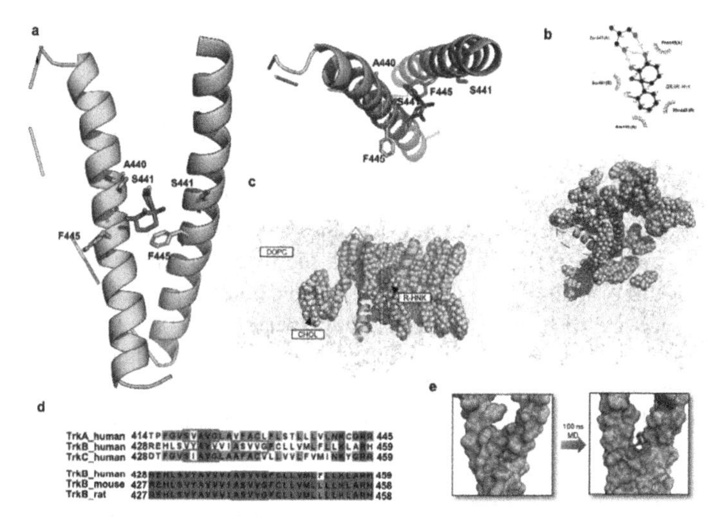

Fig. 3. AF-M predicted TrkB transmembrane helix dimer after molecular dynamics simulation to localize the drug binding pocket (A) TrkB transmembrane helical dimer with (2R, 6R)-HNK coupled structure after 100 ns MDS. Contact residues are shown in stick form. The top view is shown on the right. (B) Pose fingerprint of the (2R, 6R)-HNK contact with the surrounding TrkB transmembrane helix residues. The most stable contact occurs on Ser441, with a total of three hydrogen bonding contacts. (C) The final conformation of the TrkB transmembrane helix dimer with the (2R, 6R)-HNK-coupled structure in the lipid box, with cholesterol being displayed in a spherical shape, after 100 ns MDS simulation, the top view(right). (D) Sequence comparison of the human Trk family (top), and sequence comparison of TrkB across species (bottom), reveals that the crossover sequence is highly specific in the same family and highly conserved between species, and the helical crossover residue sequence is boxed in red. (E) After 100 ns MDS simulation, (2R, 6R)-HNK is transferred from the protein surface to the crossover gap and a drug binding pocket is generated. (Color figure online)

bonding with S441 and hydrophobic forces with V437, V438, A440, V442, and F445 (Fig. 3A). Further analysis of drug-residue interactions revealed three hydrogen bonds created between (2R, 6R)-HNK and S441 at a distance of about 3 Å (Fig. 3B). The comparing of (2R, 6R)-HNK before and after the simulation showed that the molecule moved from the protein surface to the center of the crossover, where it formed a TrkB-specific hydrophobic pocket that was exactly on the two dimer crossover sequences (Fig. 3E). This phenomenon suggested that (2R, 6R)-HNK effectuates the extracellular segment signal on the kinase segment by anchoring the transmembrane helical dimer conformation, thereby exerting a synaptic plasticity and enhancing the antidepressant effect.

3.4 AF-M Predicts Intracellular Signal Assemblies of TrkB with Multiple Binding Modes Between Kinase and Signal Proteins

- AF predicts TrkB kinase homodimers with centrosymmetry between monomers centered on the activation loop.

The kinase segments of the RTK family are structurally similar, with the main structure consisting of an N-lobe composed of α-helixes and a C-lobe composed of β-folds, as well as a hinge bridging the two globes, while a loop sequence, known as the activation loop, exists on the C-lobe and plays a critical role in the phosphorylation cascade. For TrkB, there are five majors phosphorylated tyrosines (pTyr), including Y516 in the proximal membrane region recognized by AF as an incomplete helix, Y702, Y706, Y707 in the activation loop, and Y817 in the C-terminal IDR tail. The predicted TrkB kinase segment shows a consistent conformation with the resolved kinase-inactive structure, wherein the αC helix opens outward and the expanded binding pocket appears as C-helix out; the benzene ring of Phe in the DFG at β8 appears as DFG-out towards the hinge domain, indicating that the protein is inactivated and is in a closed state. Despite the presence of the DFG-in structure of CPD5N [PDB:4AT3] with the antagonist in the database, AF2 tends to select the inactive state of the kinase for output. Among all subsequent structures of the complex, the kinase presents an inactive conformation consistent with the monomer (Fig. 4B).

Furthermore, three models of self-inhibition have been identified in studies of RTKs, including cis-inhibition of the active site by the proximal membrane region and C-terminal tail, tightening of the activation loop, and blocking of the substrate binding site. In the predicted structure of the TrkB monomer, the juxtamembrane region is identified as a disordered scattered structure, the C-terminal tail has only 10 residues and does not block the ATP-binding site, and the activation loop shows an outwardly expanded relaxed structure. Therefore, AF did not provide a monomeric TrkB self-suppressive structure.

- Predicted TrkB kinase segment complex structure with PLCγ1 and SH2 structural domain bound to the kinase C-terminus

Tyrosine kinases transmit signals downstream by recruiting and activating substrate proteins. pTyr binds to the SH2 and PTB structural domains of the signal proteins to generate cascade phosphorylation. The SH2 and PTB structural domains are present in a diverse set of proteins containing a range of catalytic type and interacting domains, which provide a degree of specificity by recognizing pTyr residues and the surrounding residues. Three extensively studied signal proteins were selected for the construction of their complex structures with TrkB kinase segments to examine the efficacy of AF-M in predicting the intracellular molecular assemblies of tyrosine kinases. Notably, the AF was previously shown to be effective in the prediction of intracellular giant complexes, with quiescent structures. However, the tyrosine kinase signal is transmitted by a liquid-liquid phase separation mechanism, and the protein structure in the droplet exhibits a high degree of dynamics.

The PLCγ signal pathway is downstream of TrkB, which plays a major role in promoting calcium inward flow. The activation of the pathway leads stimulates CaMKII and subsequent synaptic plasticity, where PLCγ1 serves as the first substrate with two tandem SH2 domains. The previous structural analysis provided two reference binding modes; one from the binding of NSH2-CSH2 to FGFR1, in which NSH2 forms a binding pocket on pTyr at the C-terminus of the kinase 31, and the other is from the complex formed by FGFR2 and CSH2. Based on this structure, another activation model was proposed as follows: CSH2 contacts both kinases simultaneously, and in addition to

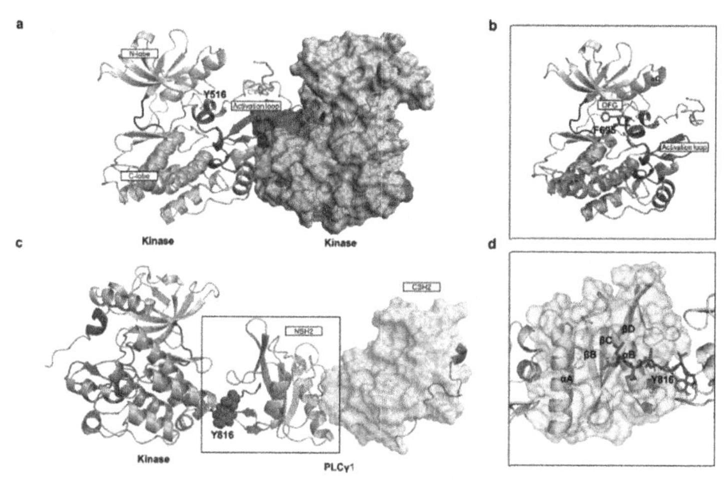

Fig. 4. AF-M predicted TrkB intracellular kinase segment homodimer, heterodimer bound to PLCγ1 NSH2-CSH2 (A) AF-M predicted homodimer of TrkB kinase segment. The two monomers are centrosymmetric with the active ring as the central point, and the active ring is close but not in contact. The proximal membrane region crosses the interface where the active ring is located. The hinge region is dark blue, and the proximal membrane region containing the autophosphorylated tyrosine, the active loop, and the C-terminal tail are purple. Same below. (B) AF-M predicted sequences related to kinase activity in TrkB kinase segment homodimers are shown. αC is C-helix out, DFG is DFG-out, and the kinase is in the off state. (C) AF-M predicted heterodimer of TrkB kinase segment with PLCγ1 NSH2-CSH2 structural domain, Y816 site in contact with NSH2, shown as spherical. (D) Details of the contact of the TrkB kinase segment with the PLCγ1 NSH2 structural domain are shown the C-terminal tail is embedded in the surface groove of NSH2 and is localized near αB. (Color figure online)

producing a pocket binding mode similar to that described above, the pTyr-containing tail at the C-terminus of CSH2 is inserted into the active groove of the other kinase. These two models provide insights into two types of kinase-substrate binding: 1. Substrate recruitment and phosphorylation are cis at the same tyrosine kinase monomer. 2. Substrate recruitment and phosphorylation are dependent on the dimerization of the kinase segment. Consequently, we selected NSH2-CSH2 of PLCγ1 and retained the pTyr-containing sequence downstream of CSH2 to construct the complexes of NSH2-CSH2 with TrkB kinase segment in a 1:1 ratio and CSH2 with kinase segment in a 1:2 ratio.

This finding indicated that AF-M predicted the signal chaperone contact domain and located the phosphorylation site in the case where the database contains the structure of the kinase complex with the signal companion. However, some differences were noted in the spatial distribution of specific residues from the resolved structure owing to sequence differences among the kinases and the additional possibilities offered by the database containing the pTyr peptide in a complex with the SH2 structural domain. Moreover, in the 1:2 complex, AF did not construct the theoretical model of transdimeric activation. CSH2 in all predicted structures was free from the dimerization of the kinase segment, indicating that despite the existence of the theoretical model, AF-M could not construct an approximate linkage due to the lack of reference structures in the database.

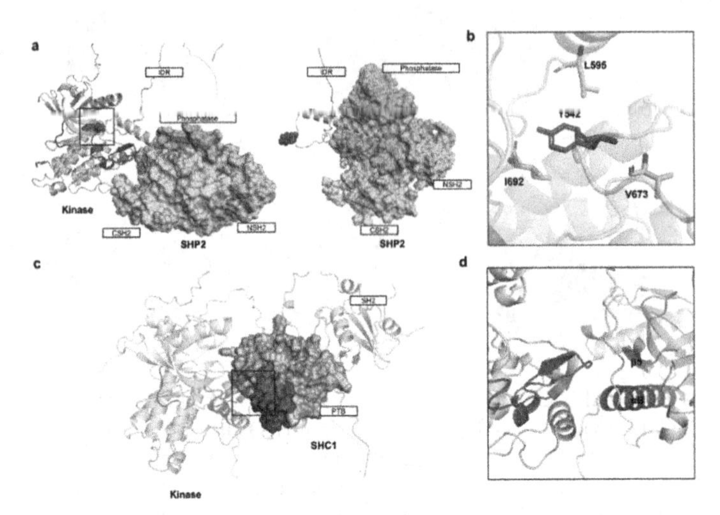

Fig. 5. AF-M predicted heterodimer of TrkB intracellular kinase segment bound to SHP2 and SHC1 (A) AF-M predicted heterodimer of TrkB intracellular kinase segment SHP2, with the structure of the disordered region of SHP2 extending into the active groove between the kinase globes. (B) Detail of AF-M predicted heterodimer of TrkB intracellular kinase segment SHP2 showing that autophosphorylated Y542 on the disordered region of SHP2 makes contact with I692, V673 and L595 of αC of C-lobe. (C) AF-M predicted heterodimerization of the TrkB intracellular kinase segment SHC1 with contact occurring in the disordered region while PTB is closer to the C-lobe compared to SH2. No contact occurs between PTB and the kinase, but the binding pocket of PTB (red) is toward the C-lobe. (D) Detail of AF-M predicted heterodimerization of the TrkB intracellular kinase segment SHC1, with β5 and αB of PTB producing binding pockets toward the C-lobe and close to the activation loop. (Color figure online)

- Prediction of the complex structure of TrkB kinase segment with SHP2 and SHC1 and identification of the different binding modes of kinase binding to SHP2 disordered region and SHC1 PTB

To further understand the effect of AF on the formation of kinase-signal protein complexes, two key signal proteins were selected to construct the dimer formed with kinase, including SHP2 and SHC1. SHP2 is an allosteric enzyme that consists of two tandem SH2 structural domains, phosphatase structural domain and C-terminal disordered tail, while NSH2 contacts with PTP and self-inhibits the phosphatase activity in the absence of biochemical reactions. In addition, SHC1 is a signal scaffold protein, and its PTB binds to the kinase structural domain, while SH2 further binds other signal proteins and large segments of the sequence show a disordered structure. Different from PLCγ1, neither SHP2 nor SHC1 obtained a resolved structure of its structural domains in a complex with the kinase segment, and the inferred contact site was derived from the contact conformation of the pTyr-containing polypeptide with SH2 and PTB.

The predicted structure of the TrkB kinase segment with SHP2 does not provide a conformation in which the PTP is activated by conformational change; however, it still adopts a self-inhibitory structure, and NSH2-CSH2 does not interact with the kinase (Fig. 5A). A phosphorylated peptide study showed that the spaced pTyr site sequentially crosses the surface of SH2 and causes NSH2 to lose the inhibition of the phosphatase; then, the kinase contacts and reacts with the main body of the phosphatase as the conformation during the allosteric activation is not available due to the absence of PTM information from AF-M. The contact in the predicted structure occurs in the C-terminal α-helix of the PTP structural domain of SHP2 and downstream to IDR, as it is embedded in the active groove of the kinase forming a multivalent and stable contact, where Y547 of SHP2, the phosphorylated site creates four hydrophobic bonds and one hydrogen bond between L595, I692, and V673 of the kinase segment (Fig. 5B). This interaction exhibits a transient local contact where SHP2 is recruited to the C-lobe as well as near the activation loop through the disordered region in the case of autoinhibition. Of the five possible conformations, four of the resulting contacts occur in the C-terminal disordered tail of SHP2. This interaction is consistent with recent studies on the liquid-liquid phase separation of SHP2, which undergoes LLPS in the autoinhibited state and is enhanced by the transformation of the phosphatase to an active conformation after mutation. Before this phenomenon, AF2-based phase separation prediction programs have been developed by scoring disordered regions, while AF-M has the potential to complete the screening and localization of phase separation targets.

In conclusion, we explored the potential conformation of AF-M in predicting the intracellular kinase segment dimerization and kinase segment binding to various key signal proteins. In the case of resolved structures with similar patterns, AF-M could accurately localize to the binding residues, such as NSH2 of PLCγ1. However, the monomer in the unresolved complex exhibits a self-inhibited structure with the blocked binding interface, and the predicted contacts occur at the IDR because the PTM information is not considered. In order to construct a complete, signal pathway network containing three-dimensional information is necessary to deduce additional conformations of kinases and companions at different reaction stages and to correlate the PTM information of the residues with these structures. On the other hand, it is essential to concatenate the previous experimental evidence and construct a three-dimensional model for artificial intelligence learning.

4 Discussion

MDD is a pressure stress-mediated physiological disorder that often leads to reduced synaptic plasticity, in which the TrkB receptor, the starting point of the synaptic plasticity pathway, is an RTK distinct from 5-HT and glutamate receptors 11. Its large extracellular receptor and intracellular kinase structures, as well as the highly variable transmembrane helices and disordered sequences between the two, have led to the resolution of its activation state being mostly limited to the extracellular segment. Following AF2 catch all protein in one draft, AF-M evolved to be able to predict complex structures, allowing us to construct TrkB dimerization activation structures as well as complex structures with intracellular signal molecules. Adopting a divide and conquer algorithm, we first predicted the activation structure of the whole extracellular segment of TrkB binding to mBDNF and were able to rapidly perform homologous ligand-receptor pairing, providing three-dimensional information that can be used to design targeted agonists or antagonists that are highly specific to a given combination. For example, the design of mBDNF-mimetic microproteins targeting TrkB reduces the aberrant activation of TrkA and TrkC.

Transmembrane helices, due to their complex environment and their role as relay stations for signal cascades, often require careful design of membrane mimics for resolution. For unresolved transmembrane helix dimers, manual screening is required among numerous docking results. AF-M provides dimeric structures that are stable in the lipid raft environment and create pockets at the crossover that bind to novel antidepressants. This will strongly facilitate the design of drugs against RTKs, breaking the inherent impression that transmembrane helical drug targets exist only for GPCRs and ion channels with multiple transmembrane helices. It will also enhance the resolution of highly flexible dimeric transmembrane helices of RTKs under Cryo-EM and refine the transmembrane dimeric activation pattern of RTKs members in combination with MDS.

As a kinase segment that functions as a biochemical reaction, its potential chaperones exceed 300 species, and the complex conformation is currently difficult to capture due to the rapidly proceeding phosphorylation reaction. Attempts were made to construct a complex pattern of TrkB kinase segment with three typical signal chaperones, and the binding site and pattern of PLCγ1 were highly similar to that of the resolved FGFR2. Unfortunately, however, the binding site of SHP2 presenting a self-repressed structure was designated as being in the disordered region due to the absence of PTM information. Rather, SHC1 successfully distinguished a binding mode with the PTB structural domain binding pTyr. Further iterations of AF-M are needed in the prediction of kinase-signal chaperones to predict the metastable binding process of proteins in biochemical reactions, such as researchers adding PTM information when uploading structures, marking key sites regulating metastable conformations, and uploading multiple dynamic conformations of the binding process to the database for artificial intelligence learning.

AF-M extends the prediction to multimers, giving us access to the underlying structures of molecular machine-protein complexes that dominate physiological functions. Such breakthroughs allow me to stand at the beginning of the next era of structural science, using deep learning-based artificial intelligence programs represented by AF2, combined with MDS, to build a dynamic signal pathway network containing structural information. In this network, the self-assembly of small molecules in all physiological and pathological processes can be demonstrated, and the operation of the entire pathway can be controlled by modifying and controlling a few key targets, which will enable drug design to shift from single-target to global. In the foreseeable future, AI, a powerful assistant, will play an important role in breaking down difficult-to-treat diseases involving multiple mechanisms, such as MDD.

References

1. Malhi, G.S., Mann, J.J.: Depression. Lancet (London, England) **392**, 2299–2312 (2018)
2. Heine, M., Holcman, D.: Asymmetry between pre- and postsynaptic transient nanodomains shapes neuronal communication. Trends Neurosci. **43**, 182–196 (2020)
3. Santos, R., Ursu, O., Gaulton, A., et al.: A comprehensive map of molecular drug targets. Nat Rev. Drug Discov. **16**, 19–34 (2017)
4. Okaty, B.W., Commons, K.G., Dymecki, S.M.: Embracing diversity in the 5-HT neuronal system. Nat Rev. Neurosci. **20**, 397–424 (2019)
5. García-Nafría, J., Nehmé, R., Edwards, P.C., Tate, C.G.: Cryo-EM structure of the serotonin 5-HT(1B) receptor coupled to heterotrimeric G(o). Nature **558**, 620–623 (2018)
6. Kim, K., Che, T., Panova, O., et al.: Structure of a hallucinogen-activated Gq-coupled 5-HT(2A) serotonin receptor. Cell **182**, 1574-1588.e1519 (2020)
7. Cao, C., Barros-Alvarez, X., Zhang, S., et al.: Signaling snapshots of a serotonin receptor activated by the prototypical psychedelic LSD. Neuron **110**, 3154–3167 e3157 (2022)
8. Kaplan, A.L., Confair, D.N., Kim, K., et al.: Bespoke library docking for 5-HT(2A) receptor agonists with antidepressant activity. Nature **610**, 582–591 (2022)
9. Cao, D., Yu, J., Wang, H., et al.: Structure-based discovery of nonhallucinogenic psychedelic analogs. Science (New York, N.Y.) **375**, 403–411 (2022)
10. Zhang, Y., Ye, F., Zhang, T., et al.: Structural basis of ketamine action on human NMDA receptors. Nature **596**, 301–305 (2021)
11. Wang, C.S., Kavalali, E.T., Monteggia, L.M.: BDNF signaling in context: from synaptic regulation to psychiatric disorders. Cell **185**, 62–76 (2022)
12. Casarotto, P.C., Girych, M., Fred, S.M., et al.: Antidepressant drugs act by directly binding to TRKB neurotrophin receptors. Cell **184**, 1299–1313 e1219 (2021)

Does Trust Improve Commercial Insurance Participation Behavior?

Lei Yu[1], Xinlong Yang[2](✉), Jue Wang[3], and Yumeng Yang[1]

[1] Xi'an University of Technology, Xi'an, China
[2] Shenzhen Institute of Information Technology, Shenzhen, China
yangxinlong@sziit.edu.cn
[3] Jiangsu University, Zhenjiang, China

Abstract. Based on data from the China Family Panel Studies micro-social survey, this paper empirically analyzes the influence of trust on commercial insurance participation. The results demonstrate that trust significantly enhances the probability and extent of commercial insurance participation. Moreover, the research findings further support the relationship between trust and commercial insurance participation. Heterogeneity analysis reveals that trust is more prominent in promoting participation among families residing in cities and towns with lower education levels. Mechanism analysis indicates that trust facilitates the purchase of commercial insurance by increasing family income. This paper's research introduces a novel perspective to the influencing factors affecting commercial insurance participation, emphasizes the importance of trust, and reinforces the development of the social credit system to foster the healthy growth of the commercial insurance market.

Keywords: Trust · Commercial Insurance Participation · Instrumental Variables

1 Introduction

In recent years, China's multi-level social security system has been continuously improved, and commercial insurance can not only play an important supplementary role in the process of social security system construction and development, but can also replace the function of social security to a considerable extent. According to the need of sustainable development of social security system, we should respect the market rules of commercial insurance, leave the corresponding space for the development of commercial insurance, and take effective policy measures to promote it (Xu 2010). In 2017, the Opinions on Accelerating the Development of Commercial Pension Insurance proposed that the development of commercial pension insurance is important for improving the multi-level pension protection system, promoting the multi-level and diversified development of the pension service industry, coping with the trend of population aging and new changes in employment patterns, further protecting and improving people's livelihood, and promoting social harmony and stability. The Opinions of the State Council of

S. Zhang et al. (Eds.): BigData 2023, LNCS 14203, pp. 166–181, 2023.
https://doi.org/10.1007/978-3-031-44725-9_12

the Central Committee of the Communist Party of China on the Key Work of Comprehensive Promotion of Rural Revitalization in 2022" points out that the state is required to increase insurance protection and optimize and improve the "insurance+futures" model. This shows that commercial insurance has become increasingly irreplaceable in meeting the diversified protection needs of the nation and maintaining social stability.

Combined with the current good development opportunities facing the commercial insurance market, the development of the existing commercial insurance market is still in need of improvement. According to the 2018 China Urban Family Wealth and Health Report, the participation rate of life insurance for households is 14.0%, health insurance for households is 10.2%, and other commercial insurance for households is 5.7%. This data shows that although some of our households are aware of participating in commercial insurance, the overall participation rate of commercial insurance is low. 2019 Swiss Re released a sigma report stating that the insurance density in China is US$430, ranking 46th in the world, with a difference of US$388 from the world average; The difference between the depth of insurance and the world average depth level is 2.93%, 4.3%, which shows that the development of commercial insurance market in China is still very lagging behind.

Since the reform and opening up, China has entered a period of social transition from a planned economy to a market economy, during which the imperfect development of the market economy and the poor quality of market traders themselves have led to a serious trust deficit problem. This problem is reflected in the insurance field: the lack of trust between insurers and policyholders, which leads to the disruption of the normal order of the insurance market and affects the normal conduct of commercial insurance transactions.

It can be seen that it is an important issue to study how to effectively increase the participation rate and involvement in commercial insurance from a trust perspective. The main contributions of this paper focus on the following three aspects: First, the existing literature has mostly studied the influencing factors of commercial insurance participation from the perspectives of household economic status, household structure and demographic characteristics, and social factors. Second, in order to obtain reliable and valid estimation results, this paper utilizes micro social survey data and uses instrumental variables to overcome the endogenous problem in the model. Third, this paper constructs a mediating effect model of trust acting on commercial insurance participation behavior through household income in order to reveal the intrinsic transmission mechanism and the path of action of commercial insurance participation.

2 Literature Review

2.1 Researches on Factors Affecting Commercial Insurance Participation Behavior

With the opening of China's financial market, commercial insurance has emerged as a crucial pillar in the country's protection system, enhancing people's well-being and improving their quality of life (Wang 2010). The development of commercial insurance serves as a valuable complement to the social security system, effectively alleviating the pressure on government social security (Zhao 2007). Currently, the existing literature on the factors influencing commercial insurance participation mostly focuses on two areas: Demographic characteristics, such as gender (Liu and Chen 2002), education level (Abu Bakar et al. 2012), lifestyle (Yadav and Sudhakar 2017), and financial literacy (Lin et al. 2017), have been identified as factors that impact commercial insurance participation behavior. Additionally, larger family sizes have been associated with a higher likelihood of insurance purchases (Dash and Im 2018).

Family economic status also plays a significant role in residents' decisions regarding commercial insurance participation. The level of household wealth, for example, influences the consumption of commercial insurance, as higher wealth levels are linked to more excellent knowledge of commercial insurance, more vital investment awareness, and a higher willingness to purchase such insurance (Yang and Liu 2019). Moreover, increasing household income has been shown to enhance the participation rate of commercial insurance (Showers and Shotick 1994). In Taiwan, the probability of purchasing insurance varies across regions, with households in northern Taiwan having higher odds of owning private insurance compared to non-northern households. Urban and town households are also more likely to have private insurance than rural villages (Liu and Chen 2002).

Furthermore, socioeconomic factors (Yadav and Sudhakar 2017), product prices, and promotion have been identified as influencers of the probability of insurance purchases (Esau 2015). He and Li (2009) noted that higher levels of social capital promote residents' insurance purchases, and increasing social interaction among residents have related to a higher propensity to purchase commercial insurance (Durlauf 2004). In Lithuania, monetary factors primarily influence insurance consumption decisions (Ulbinait et al. 2013).

2.2 Related Studies on Trust

As an indispensable part of human life, trust permeates all aspects of social life in a natural and self-explanatory form. Since the 1970s, the study of trust from the perspective of economic sociology has gradually entered the Western academia. The concept of trust is too general and richly structured, and there is still no universally accepted definition among Western economists. One of the more classic ones is Fukuyama's definition of trust in his book Trust: The Creation of Social Morality and Prosperity - based on the norms shared by members of a community and the role of individuals belonging to that community to expectations of normal, honest, and cooperative behavior among members. Other scholars have also argued that trust is the moral basis for maintaining a well-functioning market economy (Zhang and Ke 2002).

Trust, as an integral part of social capital, plays an increasingly important role in the economy and society as a whole. From a macro perspective, social trust is increasingly seen as a non-economic determinant of economic development, and its positive impact on the economic sphere of social life has been demonstrated by many studies, which is an incentive for new research initiatives examining the level of social trust, as the findings may be crucial for local policy-making.

Mularska-Kucharek and Brzeziński (2016) obtained that regions with high social trust have the highest level of development through research and analysis; high trust is a booster for trade development, low trust is a stumbling block for trade development, and trust between two countries (regions) will be beneficial to the sustainable development of bilateral trade. Chen and Qi (2022) concluded cost economics theory that the level of trust affects the size of export trade by influencing the size of transaction costs to affect the size of export trade based on transaction, with higher trust generating a trade creation effect and lower trust generating a trade barrier effect, while trading partner trust significantly affects the size of China's agricultural export trade; Bloom et al. (2012) argue that regions with high trust and strong rule of law are able to sustain large firms and industrial sectors that require decentralization. Moreover, considering the size and industry of these regions, these firms also have a higher degree of empowerment, and among subsidiaries of multinational companies, trust is important for countries with high bilateral trust, increasing the likelihood of delegation.

At the micro level, Song and Wang (2010) found that inter-firm trust and learning have a positive effect on both buyer's and seller's innovativeness through survey data of 194 mainland Chinese firms. There is a positive interaction between trust and learning. Moreover, there are interactions and complementarities between them; firms in emerging markets often face corruption and institutional weaknesses in their environment, and despite these challenges, trust can help employees to be more productive, while at the same time, firms that build trust among their employees may be more capable of dealing with the challenges posed by corruption and uncertain institutional environments (Sánchez and Lehnert 2018); Trust increases overall productivity through two channels: first, trust facilitates redistribution among firms, as CEOs can delegate more decisions, thus allowing more efficient firms to grow; second, trust complements the adoption of new technologies, thus increasing productivity technological change within firms during periods of rapid growth (Bloom et al. 2012); trust promotes inter-agent cooperation plays an important role, especially in credit lending activities. Trust building has attracted considerable research interest, and gift giving has been shown to be one of its main drivers. Through their study, Zhang et al. (2020) found that gift-giving mainly contributes to building trust at the individual level rather than at the community level. In turn, individual and community trust can facilitate access to informal and formal sources of credit, respectively. In addition, personal trust facilitates access to informal loans for consumption and medical expenses, but not for production; an increase in the level of community trust increases household risk tolerance and risk tolerance, as evidenced by a significant increase in the proportion of household financial risk assets (Zang and Wang 2017); Moderate trust maximizes household income. On the one hand, the heterogeneity of people's trustworthy beliefs, combined with individuals' tendency to infer beliefs about others from their own levels of trustworthiness, may produce a non-monotonic relationship between trust and income. Highly trustworthy individuals who believe that others are like them tend to form overly optimistic beliefs that lead them to take too many social risks, be deceived more frequently, and ultimately perform less well than those who happen to have trustworthiness levels close to the population average. On the other hand, low trustworthiness types form beliefs that are too conservative and thus avoid being cheated, but often pass up lucrative opportunities and therefore underperform China is entering an aging society where the emotional health of older adults is increasingly important and social trust is an important factor affecting the emotional health of rural older adults. Trust in family members, trust in friends, and trust in neighbors all have significant positive effects on the emotional health of older adults (Chen and Zhu 2021).

So what are the factors that affect trust? The use of the Internet can increase the level of social trust by facilitating offline socialization and improving interpersonal satisfaction, and also decrease the level of social trust by affecting users' perceptions of social justice, but the diverse information of the Internet did not decrease the level of social trust due to increased cognitive disagreement (Wang and Zhou 2019); Shi et al. (2016) found that taking educational resources as an example resource grabbing stems from the insufficient supply of public resources, and the level of distrust among people is then increased; factors such as rapidly advancing urbanization and uncertain external environment can cause a decrease in social trust (Zeng and Liu 2021).

In the process of marketization, on the one hand, the competition between people reduces the trustworthy people's expectations of humanity, and on the other hand, the market development is not yet sufficient to protect the trustworthy people, which ultimately leads to the existence of marketization's inhibiting effect on trust (Xin 2019); the transportation facilities in a certain region are perfect, on the one hand, this means that the cost of people's interactions will decrease, which will in turn promote the interpersonal interactions; on the other hand the logistics and information flow within the region and between the region and other regions will increase, which will eventually increase the trust level of people in the region; the more state agency workers in a region's population, the less trustworthy the region is, in the case of excessive power and irregular behavior of officials, the more officials in a population, the more frequent the policy changes, the more uncertain the market environment, and thus the less trust people have in that region (Zhang and Ke 2002).

Fukuyama (1995) believes that the division between low-trust and high-trust societies is based on the different cultural conditions of each region, and in his opinion our country is a typical low-trust society based on blood relations. In the process of insurance transactions, at the time of the initial signing of the insurance contract, it is actually a guaranteed promise issued by the insurer to the insured, and whether the insurer is required to fulfill its promise is not known until several years later. This unique feature of the insurance transaction determines that the insurance business is based on insurance credit, and the insurance industry cannot develop healthily if it loses its credit base.

The existing literature on the influencing factors of commercial insurance participation mainly focuses on social interaction, financial literacy, and household economic status, but there is little literature on the influence of trust on commercial insurance participation decision from the perspective of trust. Therefore, this paper analyzes the mechanism of trust and commercial insurance participation behavior from a new perspective-trust, relying on domestic authoritative databases, and provides corresponding theoretical basis and reference suggestions for the construction of social credit system and the development of commercial insurance in China.

3 Theoretical Analysis and Research Hypothesis

Based on the concept of information asymmetry, trust is beneficial to economic activities because it reduces the cost of gathering information and effectively facilitates transactions between people, between firms and firms, and between firms and individuals.

Compared with high-trust societies, low-trust societies rely more on formal institutions to ensure the enforcement of contracts, and in some countries, trust among people works to some extent as a substitute for formal institutions if the government is unwilling or unable to provide strong organizational or regulatory systems to safeguard citizens' interests (Xu 2005).

Therefore, we propose the hypothesis that

H1: Trust is an important influencing factor for commercial insurance participation. The higher the level of trust the higher the probability of commercial insurance participation and the level of participation.

The existing literature suggests that a moderate level of trust can increase the income level of families (Mo and Ye 2021), while an increase in family income will be more supportive of families' decision to participate in commercial insurance (Showers and Shotick 1994).

Therefore, we formulate the hypothesis:

H2: The increase in household income plays a facilitating role in the trust promotion process of commercial insurance participation.

4 Empirical Analysis

4.1 Data and Variables

This paper collects the data from the 2018 China Family Panel Survey Studies (CFPS), which covers 25 provinces, municipalities, and autonomous regions. The survey includes demographic, social information, household investment, and other data from more than 16,000 households, providing robust data support for studying commercial insurance participation behavior from a trust perspective.

Explanatory Variables. We have use five explanatory variables complied from the CFPS, "*How much do you trust your parents? How much do you trust your neighbors? How much do you trust strangers? How much do you trust your local government officials? How much do you trust your doctor?*" There are eleven levels of responses from very distrustful (0) to very trustful (10). This paper generates five explanatory variables accordingly: trust in parents (*TP*), trust in neighbors (*TN*), trust in strangers (*TS*), trust in cadres (*TC*), and trust in doctors (*TD*). Finally, this paper sums up the respondents' trust in five different groups of people to obtain the core explanatory variable *trust*.

Dependent Variables. This paper examines two critical explanatory variables: participation in commercial insurance and the amount of premiums spent on commercial insurance. The CFPS posed the following question regarding commercial insurance participation: "*In the past 12 months, how much did your household spend on commercial insurance (e.g., commercial health insurance, auto insurance, family property insurance, commercial life insurance, etc.)?*" In this study, we treat commercial insurance participation (*CIP*) as a binary variable, using responses more significant than 0 to indicate participation (set to 1). In contrast, responses of 0 indicate non-participation (set to 0). Furthermore, we employ the natural logarithm of the amount spent on commercial insurance to measure the extent of commercial insurance participation (*CIPE*).

Control Variables. This paper has controlled for a series of factors that influence commercial insurance participation behavior, including individual-level, household-level, and regional-level factors, such as age, gender of household head (*gender*), marital status (*married*), educations level (*edu*), hukou, physical condition (*Phys_cond*), social security (*Soc_sec*), family size(*Fmly_sz*), household debt(*debt*), urban(*urban*), net family income (*NFI*), household net worth (*HNW*), social interaction (*SI*), eastern region (*eastern*), western region (*western*), per capita GDP of the household (*GDPper*) (Butler et al. 2016; Sánchez and Lehnert 2018; Chen and Zhu 2021).

Descriptive Statistics. Based on the sample statistics presented in Table 1, it is evident that approximately 42% of the households in the sample purchased commercial insurance, while around 58% did not participate in commercial insurance. The average premium expenditure for purchasing commercial insurance is RMB 2,949 per year, accounting for approximately 3.7% of household income. The mean for trust variables in different categories are as follows: trust in parents is 9.508, trust in neighbors is 6.764, trust in strangers is 2.430, trust in cadres is 4.888, and trust in doctors is 6.574. Summing these values across the five categories yields a mean of 30.16 for the trust variable. It indicates the presence of a trust structure among the Chinese population, characterized by a "differential order pattern". Regarding demographic characteristics, the average age of household heads is 49 years, with 55.7% being male. Among the respondents, 79.3% are married, and 60.7% have rural household registration (*hukou*). 75.5% report good health, while 72.7% have social security coverage. Furthermore, 52% of the sample households have an elementary school or lower education. Regarding family size, the majority consists of three or four members. Regarding economic status, the average net household income is 79,323 yuan per year, and the average net asset value is 966,539 yuan. Among the sampled households, 63.9% reside in urban areas, and 33.5% have debts.

4.2 Model

This paper investigates the effect of trust on commercial insurance participation behavior using the Probit model, which is specified as shown in Eq. (1):

$$Y = \alpha Trust + X\beta + \varepsilon \tag{1}$$

In this paper, the dependent variables, namely commercial insurance premium expenditure, fall under limiting dependent variable. Hence, we utilize the Tobit model to estimate the impact of trust on the extent of commercial insurance participation, and the model is specified as shown in Eq. (2):

$$Y^* = \alpha Trust + X\beta + \varepsilon \qquad Y = max\left(0, Y^*\right) \tag{2}$$

Table 1. Descriptive Statistics of Variables.

Variables	Obs.	Mean	Std. Dev.	Min	Max
CIP	6,608	0.416	0.493	0	1
Ln(premium)	6,608	3.445	4.154	0	11.51
Trust	6,608	30.16	6.445	0	50
TP	6,608	9.508	1.118	0	10
TN	6,608	6.764	1.925	0	10
TS	6,608	2.430	2.202	0	10
TC	6,608	4.888	2.530	0	10
TD	6,608	6.574	2.276	0	10
Age	6,608	49.38	14.27	19	97
$Age^2/100$	6,608	26.41	14.66	3.610	94.09
Gender	6,608	0.557	0.497	0	1
Married	6,608	0.793	0.405	0	1
Edu	6,608	1.688	0.801	1	3
Hukou	6,608	0.607	0.488	0	1
Phys_cond	6,608	0.755	0.430	0	1
Soc_sec	6,608	0.727	0.445	0	1
Fmly_sz	6,608	3.509	1.815	1	17
Debt	6,608	0.335	0.472	0	1
Urban	6,608	0.639	0.480	0	1
Ln(NFI)	6,608	10.88	1.194	0	12.90
Ln(HNW)	6,608	13.02	1.299	0	16.01
Ln(SI)	6,608	7.540	1.056	0	11.00
Eastern	6,608	0.470	0.499	0	1
Western	6,608	0.208	0.406	0	1
Ln(GDPper)	6,608	20.18	0.408	19.59	21.14

5 Estimation Results

5.1 Trust and Commercial Insurance Participation

Table 2 presents the estimation results of the impact of trust on the probability of commercial insurance participation. Column (1) of Table 2 shows that trust positively affects commercial insurance participation behavior with a coefficient of 0.0055 and is significant at the 5% level. One possible explanation for these results is that the higher the level of trust in the community, the higher the trust in the market and, therefore, the higher the recognition of the products and the various services offered by commercial insurance companies and the more likely they are to purchase commercial insurance.

Columns (2)–(6) in Table 2 present the effects of trust in different groups on commercial insurance participation behavior. Among them, trust in parents and trust in neighbors demonstrate significant effects with coefficients of 0.0263 and 0.0177, respectively, both significant at the 10% and 5% levels. It can be attributed to the fact that kinship and familiarity, such as trust in parents and neighbors, help reduce the information acquisition cost associated with commercial insurance, thereby increasing the probability of participation. Furthermore, trust in strangers also significantly affects the probability of commercial insurance participation with a coefficient of 0.0242, significant at the 1% level. Initially, people tend to be cautious about strangers and may be less inclined to participate in commercial insurance based on common sense. However, in the competitive insurance market, insurance agents strive to earn customer trust by showcasing their professional expertise, successful claim cases, and attentive services. As customers become more receptive to objective evidence and exceptional skills demonstrated by these agents, trust in strangers significantly shapes the probability of commercial insurance participation. Columns (5)–(6) indicate that trust in cadres and doctors does not significantly affect the probability of commercial insurance participation. Media over-involvement and unbiased reporting increases the lack of understanding between doctors and patients and the distrust of specialists and doctors. Consequently, people's skepticism towards experts and doctors has grown, eroding the foundation of trust. Therefore, trust in doctors and trust in cadres do not significantly affect the probability of commercial insurance participation (Shen 2007).

5.2 Trust and the Degree of Commercial Insurance Participation

The estimation results of trust on the degree of commercial insurance participation are shown in Table 3. Column (1) of Table 3 indicates that trust has a significant positive effect on commercial insurance premiums with a coefficient of 0.0316 and is significant at the 10% level. One possible explanation for these results is that the higher the level of trust in the community, the higher the trust in the market and, therefore, the higher the recognition of the products and the various services offered by commercial insurance companies and the more likely they are to spend more money on insurance.

Columns (2)–(6) in Table 3 present the effects of trust in different groups on commercial insurance participation behavior. Among them, trust in parents and trust in neighbors demonstrate significant effects with coefficients of 0.0207 and 0.112, respectively, both significant at the 5% levels. It can be attributed to the fact that kinship and familiarity, such as trust in parents and neighbors, help reduce the information acquisition cost associated with commercial insurance, thereby increasing the amount of insurance purchased. Furthermore, trust in strangers also significantly affects the commercial insurance premium expenses with a coefficient of 0.140, significant at the 1% level. Initially, people tend to be cautious about strangers and may be less inclined to participate in commercial insurance based on common sense. However, in the competitive insurance market, insurance agents strive to earn customer trust by showcasing their professional expertise, successful claim cases, and attentive services. As customers become more receptive to objective evidence and exceptional skills demonstrated by these agents, trust in strangers significantly increases the amount of commercial insurance premiums. Columns (5)–(6) of Table 2 indicate that trust in cadres and doctors does not significantly affect the

Table 2. Trust and Commercial Insurance Participation

	(1)	(2)	(3)	(4)	(5)	(6)
Trust	0.0055**					
	(0.0027)					
TP		0.0263*				
		(0.0157)				
TN			0.0177**			
			(0.0089)			
TS				0.0242***		
				(0.0082)		
TC					0.0027	
					(0.0068)	
TD						0.0006
						(0.0076)
Age	0.0860***	0.0854***	0.0849***	0.0862***	0.0855***	0.0854***
	(0.0102)	(0.0102)	(0.0102)	(0.0102)	(0.0102)	(0.0102)
$Age^2/100$	−0.0992***	−0.0984***	−0.0983***	−0.0990***	−0.0987***	−0.0985***
	(0.0100)	(0.0100)	(0.0100)	(0.0100)	(0.0101)	(0.0100)
Gender	−0.0445	−0.0424	−0.0460	−0.0575	−0.0411	−0.0412
	(0.0350)	(0.0349)	(0.0351)	(0.0354)	(0.0349)	(0.0350)
Married	0.255***	0.254***	0.252***	0.257***	0.255***	0.255***
	(0.0536)	(0.0536)	(0.0536)	(0.0537)	(0.0536)	(0.0536)
Edu	0.0902***	0.0944***	0.0931***	0.0849***	0.0946***	0.0952***
	(0.0254)	(0.0252)	(0.0252)	(0.0256)	(0.0253)	(0.0252)
Hukou	0.0556	0.0603	0.0543	0.0587	0.0591	0.0593
	(0.0440)	(0.0439)	(0.0440)	(0.0439)	(0.0439)	(0.0440)
Phys_cond	0.0599	0.0642	0.0615	0.0657	0.0668	0.0682
	(0.0422)	(0.0420)	(0.0421)	(0.0420)	(0.0422)	(0.0421)
Soc_sec	−0.0148	−0.0127	−0.0126	−0.0119	−0.0128	−0.0122
	(0.0389)	(0.0389)	(0.0389)	(0.0389)	(0.0389)	(0.0389)
Fmly_sz	0.0437***	0.0443***	0.0436***	0.0449***	0.0438***	0.0438***
	(0.0111)	(0.0111)	(0.0111)	(0.0111)	(0.0111)	(0.0111)
Debt	0.238***	0.236***	0.238***	0.238***	0.237***	0.237***
	(0.0369)	(0.0369)	(0.0369)	(0.0369)	(0.0369)	(0.0369)
Urban	0.0225	0.0179	0.0197	0.0178	0.0195	0.0187
	(0.0430)	(0.0430)	(0.0430)	(0.0429)	(0.0430)	(0.0430)
Ln(NFI)	0.170***	0.171***	0.171***	0.168***	0.171***	0.171***
	(0.0328)	(0.0326)	(0.0327)	(0.0324)	(0.0327)	(0.0327)

(*continued*)

Table 2. (*continued*)

	(1)	(2)	(3)	(4)	(5)	(6)
Ln(HNW)	0.269***	0.269***	0.269***	0.267***	0.269***	0.269***
	(0.0247)	(0.0248)	(0.0247)	(0.0246)	(0.0248)	(0.0248)
Ln(SI)	0.210***	0.208***	0.210***	0.210***	0.209***	0.209***
	(0.0310)	(0.0309)	(0.0309)	(0.0309)	(0.0309)	(0.0309)
Eastern	−0.0619	−0.0642	−0.0609	−0.0627	−0.0612	−0.0612
	(0.0460)	(0.0460)	(0.0460)	(0.0460)	(0.0460)	(0.0460)
Western	−0.103**	−0.0982**	−0.0996**	−0.108**	−0.103**	−0.103**
	(0.0494)	(0.0495)	(0.0495)	(0.0494)	(0.0494)	(0.0494)
Ln(GDPper)	−0.201***	−0.194***	−0.197***	−0.205***	−0.201***	−0.201***
	(0.0635)	(0.0637)	(0.0635)	(0.0634)	(0.0635)	(0.0635)
_cons	−5.558***	−5.790***	−5.561***	−5.333***	−5.409***	−5.412***
	(1.206)	(1.224)	(1.205)	(1.200)	(1.202)	(1.203)
Pseudo R^2	0.1835	0.1834	0.1835	0.1840	0.1831	0.1831
N	6,608	6,608	6,608	6,608	6,608	6,608

Note: Standard deviations are in parentheses and ***, **, and * indicate significant at the 1%, 5%, and 10% levels, respectively. The marginal effects estimated by the Probit model are reported in the table. The same as below

probability of commercial insurance participation. Media over-involvement and unbiased reporting increases the lack of understanding between doctors and patients and the distrust of specialists and doctors. Consequently, people's skepticism towards experts and doctors has grown, eroding the foundation of trust. Trust in doctors and cadres does not significantly affect commercial insurance premium expenses.

5.3 Robustness Test

To confirm the reliability of the estimation results, we present robust estimations by the replacing model. Considering that factors such as omitted variables and two-way causality can cause bias in the estimated coefficients, making a possible endogeneity problem between trust and commercial insurance participation behavior. This section utilizes the instrumental variables approach to address potential endogeneity problems between trust and commercial insurance participation behavior. On the one hand, improved transportation in certain areas reduces interaction costs, thereby decreasing the cost of human interaction and promoting mutual trust between people. Moreover, enhanced logistics and information flow within and between regions can increase individuals' trust, ultimately improving overall regional trust (Zhang and Ke 2002). Additionally, income is a vital indicator of individual socioeconomic status. Hu (2006) noted that regions with higher per capita income tend to exhibit higher levels of trust. On the other hand, provincial transportation facilities (*TF*) and disposable income (*DI*) are not directly related to commercial insurance participation. Therefore, it is appropriate to utilize these variables as instrumental variables for analyzing commercial insurance participation behavior.

The results of the endogenous test are presented in Table 4, and we can draw three conclusions from the estimates of the models. Firstly, both instrumental variables exhibit

Table 3. Trust and Commercial Insurance Premium Expenses

	(1)	(2)	(3)	(4)	(5)	(6)
Trust	0.0316* (0.0166)					
TP		0.207** (0.0992)				
TN			0.112** (0.0557)			
TS				0.140*** (0.0496)		
TC					0.0021 (0.0424)	
TD						−0.0011 (0.0470)
Cntl_Var	Yes	Yes	Yes	Yes	Yes	Yes
_cons	−34.18*** (7.227)	−36.35*** (7.354)	−34.26*** (7.227)	−32.96*** (7.213)	−33.40*** (7.217)	−33.39*** (7.223)
Pseudo R^2	0.0757	00757	0.0757	0.0758	0.0755	0.0755
N	6,608	6,608	6,608	6,608	6,608	6,608

a significant and positive influence, with coefficients of 0.0081 and 0.576, respectively. These coefficients are significant at the 5% level, aligning with our expectations. Secondly, in the non-identifiability test, the Wald tests for endogeneity were 5.18 and 5.79, respectively, rejecting the null hypothesis of no endogeneity at the 5% confidence level. Thus, trust is significantly associated with commercial insurance participation. Moreover, the models passed the over-identified test with p-values of 0.6415 and 0.03131, respectively. These results do not reject the null hypothesis "H0: all instrumental variables are exogenous," suggesting that the instrumental variables selected in this study are indeed exogenous. In the test for weak instrument robustness, the p-values of Wald chi-square tests are 5.12 and 5.55, both significant at 5%. Therefore, the null hypothesis "H0: endogenous variables are not correlated with instrumental variables" should be rejected. These findings indicate that the instrumental variables chosen in this study do not suffer from weak instrument bias. Thirdly, after dealing with possible endogeneity problems, the trust variable still maintains a 10% significant positive influence on commercial insurance participation behavior. The result is consistent in the baseline model.

5.4 Mechanism Analysis

Further research analysis reveals that moderate trust can optimize residents' income (*Fmly_incm*) (Mo and Ye 2021). As the income level increases, more households can

Table 4. Results of Robustness Test

	(1) IV-Probit	(2) IV-Tobit
Trust	0.0881** (0.0389)	0.5670** (0.2407)
TF	0.0878*** (0.0153)	0.0878*** (0.0153)
DI	0.0001*** (0.0000)	0.0001*** (0.0000)
Cntl_Var	Yes	Yes
_cons	−8.0527*** (1.7250)	−50.5147*** (10.6846)
t ratio of TF	5.74	5.74
t ratio of DI	2.98	2.98
Wald Chi2	5.18**	5.79**
Overid	0.6415	0.3131
Weakiv	5.12**	5.55**
Pseudo R^2	0.0419	0.0419
N	6,608	6,608

support the decision of household participation in commercial insurance. This paper utilizes the Bootstrap resampling technique to test the significance of household income. This approach helps to further dissect the mechanism of the role of trust in commercial insurance participation. The test results are presented in Table 5. The role of trust in influencing the likelihood of commercial insurance participation is significant. The mediation interval for the effect of household income on trust regarding commercial insurance participation is [0.000176, 0.0005605]. Importantly, this interval does not contain 0, indicating the presence of a mediation effect, which passes the 1% significance test.

Table 5. Results of Bootstrap Mediating Effects Test- CIP

	Observed Coef.	Boostrap Std. Err.	Z	BootLLCI	BootULCI
Indirect effect	0.0003656	0.0000976	3.74	0.000176	0.0005605
Direct effect	0.00191	0.0008372	2.28	0.0002234	0.0034423
Total effect	0.0022756	0.0008348	2.73	0.0006072	0.0038087

The results data in Table 5 shows that household income plays a positive mediating role in the relationship between trust and commercial insurance participation. In other words, trust facilitates households' purchase of commercial insurance by increasing household income, thus confirming hypothesis H2. The paper proposes a potential explanation for this observation: a moderate level of trust enables households to optimize their income. As a result, households with better financial situations can more support decision-making behaviors related to participating in commercial insurance.

This paper also employs the Bootstrap resampling technique to test the significance of household income, further analyzing the mechanism of the role of trust in the degree of participation in commercial insurance. The mediation interval for the effect of household income on trust regarding commercial insurance premium expenses is [0.001633, 0.0054832]. Importantly, this interval does not contain 0, indicating the presence of a mediation effect, which passes the 1% significance test. The test results are presented in Table 6.

Table 6. Results of Bootstrap Mediating Effects Test-CIPE

	Observed Coef.	Boostrap Std. Err.	Z	BootLLCI	BootULCI
Indirect effect	0.0035202	0.0009508	3.70	0.001633	0.0054832
Direct effect	0.0163943	0.0070008	2.34	0.0030481	0.0304358
Total effect	0.0199145	0.0069953	2.85	0.0066436	0.0337174

6 Conclusions

Based on data from the China Family Panel Studies micro-social survey, this paper empirically analyzes the influence of trust on commercial insurance participation behavior, which serves as a valuable addition to the existing research on the factors influencing commercial insurance participation. Moreover, it provides a necessary reference for developing the commercial insurance market. The findings of this paper reveal two key points: firstly, an increase in trust positively influences the purchase of household commercial insurance, and secondly, higher levels of trust are associated with greater depth of residents' participation in commercial insurance. In order to deal with endogeneity issues, this paper further supports these findings by employing a two-stage instrumental variable approach.

The policy implications derived from this study are as follows: First, insurance practitioners should enhance their professionalism and cultivate a culture of integrity to bolster public trust in the industry. Second, the government should establish a social credit system rooted in morality and supported by legal measures while strengthening supervision to reduce trust violations within the insurance industry. It will foster the healthy development of China's insurance sector. Third, the government should actively shape a positive social environment by promoting and encouraging integrity-based social values.

Acknowledgement. The research of Lei Yu, Xinlong Yang, Jue Wang and Yumeng Yang is supported by the grant of The MOE (Ministry of Education in China) Liberal Arts and Social Sciences Foundation "A study on the effect of intergenerational support on household debt behavior - a perspective based on life cycle theory" [No. 18XJC790019], and Shenzhen Institute of Information Technology research start-up foundation "Research on the investment management mode of China's pension funds" [No. SZIIT2021SK041].

References

Abu Bakar, A., Regupathi, A., Aljunid, S.M., et al.: Factors affecting demand for individual health insurance in Malaysia. In: BMC Public Health, vol. 12, no. 2, p. 1. BioMed Central (2012)

Bloom, N., Sadun, R., Van Reenen, J.: The organization of firms across countries. Q. J. Econ. **127**(4), 1663–1705 (2012)

Butler, J.V., Giuliano, P., Guiso, L.: The right amount of trust. J. Eur. Econ. Assoc. **14**(5), 1155–1180 (2016)

Chen, H., Zhu, Z.: Social trust and emotional health in rural older adults in China: the mediating and moderating role of subjective well-being and subjective social status. BMC Public Health **21**, 1–13 (2021)

Chen, H., Qi, C.: The trade effect of trust: evidence from agricultural trade between China and its partners. Sustainability **14**(2), 729 (2022)

Dash, G., Im, J.: Determinants of life insurance demand: evidences from India. Asia Pac. J. Adv. Bus. Soc. Stud. **4**(2), 86–99 (2018)

Durlauf, S.N.: Neighborhood effects. In: Handbook of Regional and Urban Economics, vol. 4, pp. 2173–2242 (2004)

Esau, E.Y.R.: Factors affecting consumer purchase decision on insurance product in PT. Prudential Life Assurance Manado. Jurnal EMBA Jurnal Riset Ekonomi Manajemen Bisnis Dan Akuntansi **3**(3) 2015

Fukuyama, F.: Trust: the Social Virtues and the Creation of Prosperity. The Free Press, New York (1995)

He, X.Q., Li, T.: Social interaction, social capital and commercial insurance purchase. Financ. Res. **02**, 116–132 (2009)

Hu, R., Li, J.Y.: The composition and influencing factors of urban residents' trust. Society (06), 45–61 (2006)

Lin, C., Hsiao, Y.J., Yeh, C.Y.: Financial literacy, financial advisors, and information sources on demand for life insurance. Pac. Basin Finance J. **43**, 218–237 (2017)

Liu, T.C., Chen, C.S.: An analysis of private health insurance purchasing decisions with national health insurance in Taiwan. Soc. Sci. Med. **55**(5), 755–774 (2002)

Mo, W.Q., Ye, B.: The, "moderate" level of trust and economic income growth. Labor Econ. Res. **9**(05), 121–144 (2021)

Mularska-Kucharek, M., Brzeziński, K.: The economic dimension of social trust. Eur. Spat. Res. Policy **23**(2), 83–95 (2016)

Sánchez, C.M., Lehnert, K.: Firm-level trust in emerging markets: the moderating effect on the institutional strength-corruption relationship in Mexico and Peru. Estudios Gerenciales **34**(147), 127–138 (2018)

Shen, X.P.: Expert systems in the vision of modernity. Learn. Explor. **02**, 43–47 (2007)

Shi, Y.P., Li, X.R.: Public resources and social trust: taking compulsory education as an example. Econ. Res. **51**(05), 86–100 (2016)

Showers, V.E., Shotick, J.A.: The effects of household characteristics on demand for insurance: a tobit analysis. J. Risk Insur., 492–502 (1994)

Song, H., Wang, L.: Does inter-firm learning and trust promote firm innovativeness? Front. Bus. Res. China **4**(2), 262–282 (2010)

Ulbinaite, A., Kucinskiene, M., Le Moullec, Y.: Determinants of insurance purchase decision making in Lithuania. Inzinerine Ekonomika **24**(2), 144–159 (2013)

Wang, W.T., Zhou, J.Y.: Internet and social trust: micro evidence and influence mechanism. Finance Trade Econ. **40**(10), 111–125 (2019)

Wang, W.: The development of commercial insurance and the construction of social security system in China. Popul. Econ. **06**, 54–58 (2010)

Xin, Z.Q.: Marketization and interpersonal trust change. Adv. Psychol. Sci. **27**(12), 1951–1966 (2019)

Xu, F.Q.: The evolution and reconstruction of the relationship between commercial insurance and social security. J. Renmin Univ. China **24**(02), 95–104 (2010)

Xu, S.F.: Trust, social capital and economic performance. Learn. Explor. **05**, 222–225 (2005)

Yadav, C.S., Sudhakar, A.: Personal factors influencing purchase decision making: a study of health insurance sector in India. Bimaquest **17**(1-A) (2017)

Yang, L., Liu, Z.X.: The impact of financial literacy on household commercial insurance consumption decisions–an analysis based on the China Household Finance Survey (CHFS). Consum. Econ. **35**(05), 53–63 (2019)

Yu, J.Y.: Analysis of the causes of lagging commercial insurance in China. Enterp. Technol. Dev. **29**(07), 88–108 (2010)

Zang, R.H., Wang, Y.: Social trust and investment in risky financial assets of urban households-an empirical study based on CFPS data. J. Nanjing Audit Univ. **14**(04), 55–65 (2017)

Zeng, X.L., Liu, Z.M.: Rapid urbanization and social trust-an empirical study based on comprehensive social survey data in China. Contemp. Finance Econ. **07**, 13–23 (2021)

Zhang, T., Liu, H., Liang, P.: Social trust formation and credit accessibility—evidence from rural households in China. Sustainability **12**(2), 667 (2020)

Zhang, W.Y., Ke, R.Z.: Trust and its explanation: a cross-province survey analysis from China. Econ. Res. **10**, 59–70 (2002)

Zhao, X.Z.: The role of commercial insurance in the social security system. Financ. Econ. **04**, 17–18 (2007)

Research on Big Data Empowering Ecological Governance

Xuejiao Liu[1] and Yi Li[2(✉)]

[1] School of Philosophy, The Institute of State Governance, Huazhong University of Science and Technology, Wuhan 430074, Hubei, China

[2] School of Marxism, Shenzhen Institute of Information Technology, Shenzhen 518172, Guangdong, China

liyiify@126.com

Abstract. The application of big data in the field of ecological governance has promoted the development and growth of intelligent ecological governance. Scientific discoveries enhance human ability to understand nature, and technological inventions enhance human ability to transform nature, which provide solid theoretical support for big data empowering ecological governance. In practice, big data technology can help to control existing pollution and prevent pollution from occurring. Big data also plays an important role in ecological restoration. Its prominent impact on soil restoration and biodiversity conservation fully demonstrates the enormous potential of big data in serving ecological governance. In addition, big data technology can help realize the economical use of resources and promote sustainable development. With the continuous improvement of the application of big data technology, the governance ability of ecological governance body has been greatly enhanced, which significantly improves the efficiency of ecological governance. Big data has enhanced the coordination and interaction among various bodies and improved the level of ecological governance on the whole. The integration of big data with other high and new technologies has promoted the expansion of the application scope of big data. All these demonstrate the important role of big data in promoting the modernization of ecological governance. A series of practical cases collected and listed in this paper provide strong proof for the use of big data in serving ecological governance. In conclusion, in the information age, people should vigorously promote the application of big data in the field of ecological governance, empower ecological governance with big data, and enable information technology to serve ecological civilization construction.

Keywords: big data · service · ecological governance

1 Introduction

The application of big data in the field of governance has directly promoted the emergence and development of intelligent governance. Intelligent governance refers to the continuous state and process in which intelligent technology means are relied upon and utilized, under the guidance of public authorities, and with the active participation of

S. Zhang et al. (Eds.): BigData 2023, LNCS 14203, pp. 182–191, 2023.
https://doi.org/10.1007/978-3-031-44725-9_13

market bodies, social bodies and individuals, to jointly reduce the cost of public affairs, improve the efficiency of public affairs and optimize the experience of public affairs [1]. Intelligent governance emphasizes guiding the governance of public affairs with a digital way of thinking, and constantly improving the institutional guarantee of laws and regulations to enable the digitization of public governance through the empowerment of digital technology.

Intelligent governance in the field of ecology and environmental protection emphasizes the penetration of intelligent governance in the field of ecology and environment. It advocates the transformation of traditional governance approaches through the wide participation of multiple bodies under the government's leadership and the application of information technology in ecological governance to achieve the digitization and intelligence of ecological governance. Ecological intelligent governance highlights the potential application of big data in ecological and environmental protection, and it is a practical case of big data empowering ecological protection.

2 Theoretical Foundation

As a complex technology tool, big data has been applied to the broad field of social governance. As an emerging technology, it has been widely used in ecological governance, forming a new trend of precise positioning, scientific analysis and effective management of environmental pollution. The positive role of science and technology in ecological governance provides a solid theoretical foundation for big data to enhance ecological governance.

2.1 The Role of Scientific Discovery

'Science, in its broadest sense, refers to the theoretical knowledge that guides human interaction with external things, and usually, above all, it refers to theoretical knowledge that guides human interaction with the nature' [2]. In this paper, science is defined in a narrow sense, namely, science is the general term of human activities to consciously understand nature and explore the unknown world.

From the perspective of the development history of science, scientific discoveries have solved a series of natural mysteries, helping human beings to understand nature and understand its laws. From the perspective of the history of science, the forms of science mainly include natural history, mathematical experiment science and rational science. Natural history observes, describes and classifies plants, animals and ecosystems from a macro perspective, helping people to understand nature as a whole and to comprehend its richness and diversity. Natural history science enhances human understanding of the natural world and helps human comprehend nature. It is a bridge of communication between nature and human. It is of great significance to alleviate the tension between human and nature that has emerged since modern times.

After the 17th century, modern experimental science of mathematics began to rise and develop. This was the time when science entered the research phase and scientists began to study the causes and formation of natural things. After more than 300 years of development, modern mathematical experimental science has formed a relatively complete subject category by the end of the 19th century, and science has been greatly improved. With the improvement of human observation, measurement and other technological tools, human beings have continuously gained batches of scientific achievements in various fields. Scientific research related to the study of nature is becoming more comprehensive and in-depth, greatly enhancing human understanding of nature.

Greek rational science proposed that the nature of the world is 'logos', which has purity and rationality. It pursues science itself and forms a noble scientific spirit. Greek rational science yielded abundant achievements, advocating to use human reason to understand the universe, and paving a way for human to explore and understand the laws of nature. Although Greek rational science does not pursue the practicability of science or pay attention to the its application, it has played a significant role in human understanding of nature.

In conclusion, scientific discoveries have continuously enhanced human understanding of the laws of nature, which makes it possible for human to avoid greater environmental pollution and ecological damage.

2.2 The Role of Technological Inventions

Technology has influenced and changed human life in various ways. It has been around since the birth of humankind. In history, there were three major technological revolutions that triggered industrial revolution and then had epoch-making effects due to technological progress.

In the mid-18th century, Watt made an improved steam engine. The invention of steam engine ended the manual labor period in which human beings mainly relied on manpower for two million years, and modern machine production began to become popular. Machines replaced human labor. Human began to get rid of the limitations of natural forces, and productivity was greatly improved. At the same time, production relations were adjusted, and the field of transportation was also changed. This industrial revolution fundamentally changed the cognitive structure of human beings and the direction of social development.

In the 19th century, the capitalist economy developed rapidly and modern scientific discoveries emerged one after another. Since the mid-19th century, scientific discoveries have been closely integrated with industrial production, and the deep combination of science and technology has produced a large number of technological inventions. Electrical appliances were widely used, which promoted the improvement of productivity again, and human beings entered the 'electric age'. The power, chemical, petroleum, automobile and other industries that emerged in this industrial revolution have improved people's quality of life in terms of energy utilization, transportation and other aspects, and mankind's ability to use technology to transform nature and create a suitable living environment has been continuously enhanced.

In the 20th century, due to the development of communication, the emergence of radio, radar and signal detection technology, the speed of information generation and transmission soared, and there are more and more means of information transmission, and human beings began to enter the information age. The core technology of the third technological revolution is electronic computer technology. The generation and development of electronic computer has greatly improved the computing speed of human beings and replaced the mental work of human beings. Nowadays, it can also partially simulate the intelligent activities of human beings, which has fundamentally changed the development process of modern society and driven the development of a large number of high-tech technologies. Automation, intelligence, information, digitization and so on have become the basic characteristics of the information age.

The three technological revolutions have had an unprecedented impact on the process of human development. And the emergence of each technological revolution has made great breakthroughs in the production capacity of human beings. Behind the technological revolution is the great progress in human understanding of nature and laws of nature. Before the British Industrial Revolution, technology mainly came from the summary of people's daily production and life experience, and the technology in this period was mainly empirical technology. After the Industrial Revolution, modern science developed rapidly, and a series of new scientific discoveries promoted the progress of technology. Technological invention was mainly the application of science. The type of technology in this period changed from empirical technology to scientific technology.

On the basis of scientific understanding, human beings' ability to utilize nature and transform nature is constantly improving. Human beings are becoming better at using technology to achieve the goal of creating a better life for human beings. Technology is playing an increasingly significant role in enhancing human beings' ability to transform nature. With the help of science and technology, human beings can realize the planned adjustment and control of human's transformation activities to nature, at the same time, they use science and technology to eliminate the harm of human activities to the ecology and environment, maintain the harmonious relationship between human and nature, so as to protect the balance of the entire ecosystem.

3 The Practice and Innovation

In the information age, the development of advanced technologies such as big data, cloud computing, the Internet of Things, artificial intelligence, 5G and other high-tech technologies has played a comprehensive role in promoting the development of human society and even the transformation of social interaction methods, and they penetrated into every corner of human life [3]. In practice, People are constantly trying to apply digital and intelligent technologies to specific pollution control, ecological restoration and environmental governance, demonstrating the bright prospect of ecological intelligent governance.

3.1 Empowering Pollution Control

Environmental pollution control is an extremely important work. The quality of the environment is not only directly affecting people's lives and health, but also closely related to public well-being. Zhi Hua X. et al. [4] analyzed the impact of air pollution and water pollution on well-being by matching provincial pollution data and individual well-being data. And the study showed that the increase of nitrogen dioxide concentration and wastewater discharge would significantly reduce personal well-being. Through model-based empirical analysis, Junjun Zheng et al. [5] concluded that with the improvement of material living conditions, people's environmental protection concept gradually strengthens, and they become more proactive in paying attention to and advocating for a green lifestyle. In the future, residents will be more sensitive to environmental pollution, and their well-being and even their quality-of-life level will be closely related to environmental quality. George MacKerron [6] conducted a survey of more than 20,000 British participants using smartphone and found that respondents were happier in all types of green environments or outdoor activities than in urban environments, indicating the positive impact of beautiful natural environments on individual well-being.

Big data is beneficial for managing existing pollution. To address the existing pollution control, first of all, it is necessary to have a clear understanding of the current situation of pollution and obtain various information about environmental pollution. Environmental monitoring technology plays an important role in understanding pollution information. It can help us timely and accurately understand changes in environmental quality and various indicators, grasp the overall state of environment, and provide scientific and effective basis for environmental management, pollution source control and environmental planning.

Big data assists in pollution prevention. The application of information technology provides data support for the delineation of pollution areas and scientifically planning of pollution prevention and control measures, making various plans for pollution prevention and control more targeted and operational. It helps to timely curbing the spread of pollution and reducing the damage to a minimum. Nowadays, environmental monitoring technology has been effectively used in various pollution control measures, demonstrating the significant effectiveness of information technology in environmental protection.

In the aspect of urban air quality monitoring, information technology can be used to make real-time disclosure of the concentrations of various pollutants in the atmosphere, so as to realize the transparency of information. Citizens can use mobile devices and other tools online to check the local air quality of the day, in case of severe air pollution, they can timely and accurately report to the relevant departments, better playing the role of public supervision. Monitoring technologies supported by big data also play a significant role in monitoring vehicle exhaust emissions and enterprise exhaust emissions. In addition, monitoring technologies are also widely used in water quality monitoring, radioactive source monitoring and many other aspects, greatly facilitating the work of wastewater treatment and the management of radioactive substances.

3.2 Empowering Ecological Restoration

As an important measure to improve ecological quality, ecological restoration is an artificial restoration activity aimed at the damage of ecosystem structure, functions and other problems. With the help of ecological restoration, partial or complete restoration of the ecosystem can be realized, so as to promote its sustainable development. The practical effectiveness of information technology in ecological restoration can be seen from the practice of ecological restoration of contaminated soil and intelligent management of biodiversity conservation.

Big data helps soil remediation. Soil pollution has hidden characteristics and is difficult to be detected, making the work of soil remediation challenging. The application of information technology greatly facilitates the monitoring of soil quality, enabling the timely identification of contaminated areas and the early implementation of pollution control, reducing the alleviation of the difficulty of cumulative pollution control. Information technology also provides convenience in understanding the current situation of polluted land, such as soil erosion, desertification, and saline-alkali land, which reduces the expenditure of manpower, resources, and finances, thereby lowering the cost of remediation. It took Da'an City of Jilin Province in China 6 years to find out the intelligent plan of saline-alkali land improvement, and successfully realized the application of information technology in saline-alkali land restoration, which demonstrates the broad prospects of big data technology [7].

Big data contributes to biodiversity restoration. As the foundation of life on earth, biodiversity provides crucial ecological support for human survival and development. However, currently, 'the rate of species loss has accelerated approximately from one species per day to one species per hour'. The application of information technology in biodiversity conservation is of great practical significance and far-reaching future significance. The digitization of biodiversity conservation has evolved from basic video monitoring techniques to the comprehensive utilization of intelligent use of big data, artificial intelligence, and the Internet of Things. It Has injected powerful scientific and technological force into biodiversity conservation. With the help of the biodiversity big data platforms, interdisciplinary integration of biodiversity data such as genes, species and ecology has become possible. In-depth utilization of data can be realized on the basis of shared data, providing scientific data support for research at various levels.

In a specific case practice, Houkun Hu, the rotating chairman of Huawei in 2022, introduced a case of the application of information technology in the conservation of gibbons, an endangered animal. In the conservation work of gibbons, optical video monitoring technology used to track gibbons attracted their attention, and then gibbons would destroy the monitoring tools, rendering the monitoring ineffective. Inspired by big data, the researchers combined the characteristics that gibbons are good at singing and different kinds of calls are very different, and switched to acoustic technology. By utilizing artificial intelligence sound monitoring system, cloud recording and other methods, they created a unique acoustic identity card for each gibbon. This approach effectively improved the tracking and monitoring of gibbons and facilitated timely protection in unexpected situations.

The protection of rare species is only a small aspect of biodiversity conservation. Information technology provides a solid scientific and technological foundation for ecological restoration in various fields, such as forest and grassland ecological restoration, territorial and spatial ecological restoration, biosynthesis research and so on.

3.3 Empowering Resource Conservation and Sustainable Development

Sustainable development emphasizes that mankind should adhere to a development model that meets the needs of the present without compromising the needs of future generations to meet their own needs. Information technology is the most environmentally friendly and green technology, with virtually limitless capabilities. This is mainly due to the low resource consumption and minimal pollution generated by the information industry, making it the industry that best conforms to and adapts to the requirements of green development. The information industry has become an important driving force for promoting green development. In addition, the integration of information technology into other industries is conducive to saving resources and improving the utilization rate of resources, which is of great significance for sustainable development. For example, numerous cases such as paperless office and intelligent garbage sorting and recycling demonstrate the positive effect of information technology on sustainable development.

4 The Modernization of Ecological Governance

Humanity is gradually entering a new era of data which not only affect people's behavior patterns but also significantly changes their thinking patterns. In the data era, people shift from making judgments based on intuition and experience to making comprehensive decisions based on data and analysis, significantly enhancing the scientific and effectiveness of decision-making. The practical effects of information technology in ecological governance at present, especially the positive effects in pollution control, ecological restoration, resource conservation, improving the ability of environmental situation prediction and comprehensive decision-making, etc., have already laid the practical foundation of ecological intelligent governance.

4.1 Enhancing the Capacity and Improving the Efficiency

Taking the pollution control of enterprises by environmental protection departments as an example, enterprises are the main body of the market, and effective pollution control is an important task for environmental protection departments. Traditional methods of relying on manual supervision to control enterprise pollution are not effective in truly pollution supervision and are not helpful in curbing unauthorized emissions or excessive emissions.

By actively adopting information technology in pollution control, real-time online monitoring of enterprise pollution can be achieved, greatly improving efficiency. Big data technology enables the automation and intelligence of enterprise pollution monitoring work. Environmental protection workers can accurately and timely understand the situation of pollution emissions without leaving their homes, master various emission information of enterprises, and give timely warnings of excessive emissions. It can effectively curb the occurrence of excessive and unauthorized emissions, and significantly improve the effectiveness and scientific nature of pollution control.

In addition, information technology has also achieved remarkable results in strengthening the environmental risk warning capabilities of management departments, emergency response capabilities for sudden environmental incidents and predictive capabilities for environmental situations. In response to sudden environmental incidents, digital video technology can be used to remotely understand the situation of the scene, timely conduct emergency command video conference, and in this way it can limit the harm caused by environmental events to the minimum range.

4.2 Improving the Level of Ecological Governance

The generation and resolution of ecological and environmental problems are extremely complex. 'The public nature of environmental problems is crucial to environmental problems. Their causes are decentralized and interconnected, and the harm and impact they produce are extensive, cumulative, and persistent.' [9]. It is no longer possible to rely solely on any one party to deal with the increasingly complex ecological issues. It is necessary to continuously strengthen the comprehensive cooperation among various stakeholders, achieve the cooperative governance of ecological and environmental issues, and finally enhance the level of governance and its effectiveness.

The development of information technology greatly facilitates the cooperation among various governance entities. In terms of the cooperation between ecological and environmental protection departments, information technology can not only standardize the basic data and basic business of environmental protection business, but also enhance the effective decomposition and integration of the departmental business, enabling the comprehensive cooperation between departments.

In addition, the establishment of the big data service platform for ecological and environmental protection gathers the business data from of environmental protection systems in various regions. With the support of the Internet and the Internet of Things, the intelligent analysis of data is carried out to continuously build and improve the information system for environmental monitoring, management and public services. On the platform, all bodies can obtain real-time information on various aspects of environmental quality, such as air, water and other aspects, and the public can report and complain about pollution incidents to effectively protect their ecological rights and interests. The environmental protection department can grasp the environmental situations at any time, promptly detect anomalies, and facilitate dynamic management.

4.3 The Integration with Other High-Tech Technologies

The application of information technology itself in ecological governance has greatly improved the current ecological governance ability and ecological governance level, and provided a realistic possibility for ecological intelligent governance. In addition, the integrated development of information technology and other high and new technologies has also injected powerful scientific and technological forces into ecological governance, playing an important role together in ecological governance.

Significantly, the fusion of information technology and biotechnology stands out. Biotechnology is characterized by low cost, simple operation and sustainability, so it occupies an important position in ecology and environmental protection. The integrated development of information technology and biotechnology has provided new impetus for ecological governance. The integration and development of biotechnology and information technology will promote the research and development of advanced biological manufacturing. Relying on biological manufacturing technology, we can try to replace chemical raw materials and processes with biotechnology, develop high-performance biological environmental protection materials and biological agents, promote the deep integration of chemical, material and other industrial product manufacturing with biotechnology. To achieve green, low-carbon, non-toxic, low-toxicity and sustainable development. At the same time, biological technologies such as microorganisms and enzyme preparations will be used to solve environmental pollution problems such as phosphorus removal in water bodies, heavy metal soil remediation, and waste plastic utilization and disposal, so as to facilitate the smooth progress of pollution prevention and control.

5 Conclusion

The development of science and technology is of great assistance to human understanding and transformation of nature, especially with the promotion and application of green technology in recent years. Green technology is developed to solve ecological and environmental problems, which mainly includes two types of technology: protection of green technology and promotion of green development. For example, sand control technology and sewage treatment technology belong to protection of green technology, while efficient utilization of solar energy technology and development of new energy technology arise for promotion of green development. The direct effect of green science and technology is reflected in the effectiveness of current pollution control. In the long run, green science and technology will bring about fundamental changes in human production and life style, and fundamentally realize sustainable development and ecological protection.

It is worth noting that the realization, application and promotion of green technology involve not only the feasibility of science and technology, but also the support of economic, social and policies. In reality, there are many scientifically correct and technically feasible pollution control technologies and ecological agriculture technologies, fail to achieve sustainable application due to the lack of necessary economic policy support or low profitability and public participation. Therefore, in order to play the positive supporting role of science and technology in ecological governance, it is necessary to take into account various factors and overcome many obstacles. In particular, the government should actively promote various scientific research and technological inventions related to ecological governance, provide policy support for ecological intelligent governance, and become a proactive promoter and strong supporter of ecological intelligent governance, allowing green science and technology to inject powerful scientific and technological force into ecological governance, and play a supportive role in ecological intelligent governance.

References

1. Changbo, F.: Comprehensively promote intelligent governance and create a new era of good governance. J. Natl. Sch. Adm. **02**, 59–63 (2022)
2. Guosheng, W.: What is science. China Econ. Rep. **10**, 115–117 (2014)
3. Jixi, G.: Theoretical logic and practical path of digital transformation of ecological governance. Gov. Res. **3**, 33–41 (2020)
4. Zhihua, X., Xiangang, Z., Hui yi, Y., Ying, Q.: Research on the impact and pricing of environmental pollution from the perspective of public happiness. J. Chongqing Univ. (Soc. Sci. Ed.) (4) (2018)
5. Junjun, Z., Can, L., Chengzhi, L.: The Impact of environmental pollution on the well-being of Chinese residents: an empirical analysis based on CGSS. J. Wuhan Univ. (Philos. Soc. Sci. Ed.) **04**, 66–73 (2015)

6. MacKerron, G., Mourato, S.: Happiness is greater in natural environments. Glob. Environ. Change (2013)
7. 'Alkali Bala' Bloom New Hope. - Da'an City Technology Helps Improve Saline- alkali Land. http://www.jl.gov.cn/szfzt/jlssxsxnyxdh/gddt/202209/t20220913_8573153.html. Accessed 3 Apr 2023
8. Fengchun, Z., Wenguo, Z.: Interpretation of biodiversity (Part 1): concepts and current status. Environ. Prot. **9**, 45–48 (2010)
9. Wei, X.: Starting from the 'prisoner's dilemma': a methodological discussion of global environmental issues. Philos. Res. (1) (1999)

Research on Complex Financial Ecosystem Modeling and Traceability Analysis Based on Heterogeneous Graph Attention Networks

Xiaochen Liu(✉)

China Everbright Postdoctoral Research Center, China Everbright Group, Beijing, China
ann4498@sina.com

Abstract. This paper proposes a complex financial ecosystem construction model based on heterogeneous graphical attention networks, which is based on the textual features as well as attribute features of each financial service entity node obtained in the previous section to be processed separately. And then proposes a risk propagation model based on the complex financial ecosystem network and the corresponding traceability mechanism, which is based on the previously constructed complex financial ecosystem network structure, analyzes the influence of different network structure characteristics on the risk propagation range of financial entities, so as to provide a basis for the reliability of the complex financial ecosystem network as well as risk prevention and control analysis. By constructing the financial risk propagation model and analyzing the risk propagation range, the risk control level can be upgraded for the nodes of financial entities with a larger risk propagation range so as to increase the reliability of the whole financial ecosystem network.

Keywords: Complex financial ecosystem · Heterogeneous Graph Attention Networks · Traceability

1 Introduction

The wave of digitization sweeping the globe has posed challenges to the security of financial institutions around the world. The governments emphasizes that the use of digital technology to enhance the risk prevention and control capabilities of financial institutions is an important part of building new advantages in the digital economy; the Financial Regulatory Authority have successively introduced policies related to financial technology and digital transformation to make clear the core position of risk prevention and control, pointing out that we should make good use of big data, artificial intelligence and other technologies to realize the Risk of early identification, early warning, early disposal, and enhance the ability of financial risk prevention and control. At the same time, the U.S. National Economic Council issued a white paper on "Financial Technology Regulatory Framework", especially mentioning the need to vigorously promote the application of artificial intelligence technology for the prevention and control

S. Zhang et al. (Eds.): BigData 2023, LNCS 14203, pp. 192–205, 2023.
https://doi.org/10.1007/978-3-031-44725-9_14

of risks in the financial industry. As a developed economy in Asia, Singapore has also set up a special "Financial Technology Agency" to promote the development of digital technology to enhance the risk prevention and control capabilities of financial institutions. Therefore, the integration of digital technology with the traditional business of the financial industry, and the enhancement of competitiveness, security, risk prevention and control capabilities are important issues that need to be urgently addressed by financial institutions in various countries.

Financial control group as the development goal of global large financial institutions, which can promote the classification, integration, planning and deployment of resources through the synergy of various subsidiaries, give full play to the business characteristics of mixed operations and all-around financial group, to meet the diversified needs of customers. However, business diversification, close linkage and complexity of hierarchical structure also bring new challenges to the accountability mechanism and risk control of the group. Theoretically, the financial control model is not only conducive to the effective isolation of financial and industrial segments, inhibiting internal transactions and preventing risk transfer, but also better able to play a synergistic effect and enhance the ability of diversified financial services. In practice, it is often difficult to grasp the "degree" between risk segregation and synergistic effect, and the scope of potential risks is extremely underestimated. Therefore, how to realize effective risk management and control of financial control groups is a hot issue that scholars at home and abroad are concerned about.

Financial control risk management research focus on institutional mechanism optimization, management strategy research, and gradually in the financial sector risk monitoring, early warning, risk model, control, dissemination and other aspects of the attempt to explore, but for the financial control group as a whole, the risk of accountability for less research; foreign scholars have focused on the digital era of the financial environment of the dynamic complexity of the problem of research, the use of mathematical derivation, scholars have been focusing on the dynamic complexity of the financial environment in the digital era, using mathematical derivation, computer technology and other cross-disciplinary solutions to the risk management problems of financial institutions, but the financial ecology of large-scale gold-controlled groups and the corresponding risk propagation model is still in need of in-depth research. The complex financial ecosystem of financial control groups has similarities with the ecological evolution model of the natural world. Drawing on the ecological evolution model and based on the principle of "complex network", the study of the topology and evolution process of the complex financial ecosystem network of financial control groups provides a new research idea on the business synergy and risk management of financial control groups. This project will focus on improving the risk prevention and control ability in the financial field, solving the problems of weak risk prevention ability and untimely disposal in the existing financial field through big data, artificial intelligence and cloud computing, proposing the risk accountability method based on the complex financial ecosystem, accelerating the penetration of financial science and technology risk control in multiple scenarios, and aiming to take financial risk control and accountability as the core competition to improve the safety of the entire financial industry of the group.

This paper proposes a complex financial ecosystem construction model based on heterogeneous graphical attention networks, which is based on the textual features as well as attribute features of each financial service entity node obtained in the previous section to be processed separately: for the attribute features, they are converted to vectors directly through One-Hot coding; for the textual features of the nodes, the improved Service-Text Rank approach is utilized to For node text features, the redundant content is removed, and then the text is converted to vectors by Doc2vec method. This method can realize adaptive learning of the feature vector of each financial entity node in complex financial ecosystem, and input the text embedded value and attribute embedded value into the multilayer perceptron network after splicing them together to calculate the probability of establishing links between each node, so as to get the spatial structure of complex ecosystem network to construct the whole financial ecosystem. The model can analyze the semantics of different meta-paths in the complex financial ecosystem network, select the neighboring entity nodes that have an association relationship with each financial service entity, and use the heterogeneous graph attention network to learn the feature representation of these associated nodes, and get the feature embedding values by aggregating the text and attributes of the neighboring nodes in order to enhance the prediction accuracy of the links between the nodes, so as to construct a complex financial ecosystem Model.

2 Predictive Model for Linking Complex Financial Ecosystems Based on Graph Neural Networks

With the rapid development of digital technology, the concepts of API economy, open banking and open financial control are more familiar to everyone. According to the definition of the Boston Consulting Group (BCG), open banking refers to: "In order to comply with the trend of integration between banking platforms and third-party platforms, customer demand-oriented, ecological scenarios as a touch point, API (Application Programming Interface) or integrated multi-functional combination of services (Mashup) and other technologies as a means to fragmentation of services, data commercialization, characterized by the integration with third-party data, algorithms, business, processes, etc., to achieve business-driven application architecture transformation from a business-driven to a business-driven. Characterized by fragmentation of services and commercialization of data, through integration with third-party data, algorithms, business and processes, it realizes business-driven transformation of application architecture and upgrading of the overall system from the front office to the back office, thus turning into a new-age bank. The basic business involved in open financial control is even richer, and in addition to banking interfaces, it can also provide services and data interfaces for securities, insurance, trusts, funds, and even industrial segments, which can realize greater economies of scale and economies of scope than those of open banks. Therefore, based on the open financial control scenario, the complex financial ecosystem model can be established by portraying the different correlations between its APIs and Mashups, which is conducive to suppressing insider trading and preventing risk transfer from the theoretical basis; and can also be linked with different combined services to better utilize synergistic effects and enhance the ability of diversified financial services.

In this paper, we use graph neural network based to predict links in complex financial ecosystems, applying graph neural network to the calculation of attribute embedding values and text embedding values, so as to learn the influence of neighboring service nodes on links, and designing a service link prediction model (WSLG) based on graph neural network as shown in Fig. 1.

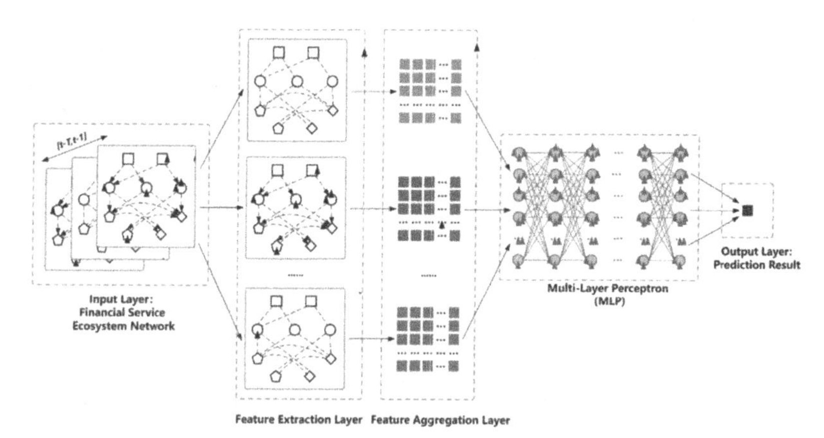

Fig. 1. Prediction of financial ecosystem links based on heterogeneous graph attention networks

Input Layer: The financial ecosystem network $SNet = (V, E)$ is used as input data to the model along with information about the various types of nodes themselves.

Feature Extraction Layer: The feature extraction layer is responsible for processing the attribute values and functional description text of the nodes in the financial ecosystem. Link prediction is also known as link embedding and feature extraction can be considered as the most important part. Attribute and functional description texts of nodes will be processed separately, and the calculation of embedding values will be performed based on graph neural network by utilizing the interaction relationship in the service network. For the attribute embedding of different financial services, in addition to the predicted attributes of the target Mashup and the API service itself, it is necessary to find out all the neighboring nodes that have an association relationship with the target financial service link, and realize the attribute embedding of the associated nodes by aggregating their representations through the graph neural network. The embedding of functional description text, on the other hand, uses the algorithms in natural language processing to achieve semantic embedding of the text, and then expands the original text into a generalized functional description text based on the graph neural network approach by means of the neighboring services that can embody the service usage scenarios.

Feature Aggregation Layer: The Feature Aggregation Layer aggregates the attribute feature embedding values and text feature embedding values of the target financial ecosystem node links output from the Feature Extraction Layer through the Multilayer Perceptron (MLP) to calculate the final prediction results, as shown in Eq. (1).

$$\hat{p}_{ij} = \text{Sigmoid}\left(MLP\left(Z_i^{Attr} \oplus Z_j^{Attr} \oplus Z_i^{Text} \oplus Z_j^{Text}\right)\right) \tag{1}$$

where $\hat{p}_{ij}$ denotes the Mashup service M_i and the API service A_j the possibility of establishing a link between the Mashup service and the API service, the Z_i^{Attr} and Z_j^{Attr} denote their attribute embedding values, respectively, and Z_i^{Text} and Z_j^{Text} denote their text embedding values, respectively, and $\oplus$ denotes the splicing operation. The embedded values are spliced and aggregated by MLP network, and then the Sigmoid function is used to control the result in the range from 0 to 1.

Output Layer: Outputs the prediction result $\hat{p}_{ij}$. The loss function is mainly used to calculate the difference between the predicted value and the actual value, and the smaller the result indicates the better result. Since the link prediction problem is similar to the binary classification problem, and the classification algorithms commonly use cross entropy as the loss function. The calculation of cross entropy loss is shown in Eq. (2).

$$\text{Loss} = -\frac{1}{N}\sum\nolimits_{M_i \in V^{\text{Mashup}}, A_j \in V^{API}} p_{ij}\log\hat{p}_{ij} + \left(1 - p_{ij}\right)\log\left(1 - \hat{p}_{ij}\right) \quad (2)$$

where, N denotes the number of Mashup and API node pairs. The Adam algorithm will be used to minimize the loss, which differs from the stochastic gradient algorithm in that its learning rate is dynamic. By calculating the first and second order moment estimates of the gradient and then adaptively updating the learning rate for each parameter, the Adam algorithm is able to achieve a faster error descent process.

3 Linked Node Attribute Embedding Algorithm

In neural networks, the attention mechanism is designed so that it can select specific inputs, which gives it the property of being able to focus more on a subset of features. Graph Attention Networks can handle inputs of arbitrary size, thus focusing on the most important part of a computational process, with the advantages of efficiency, flexibility and portability. The number of associated nodes for each target financial service is not fixed, and the graph attention network is just able to accept inputs of different sizes, using the attention mechanism can also assign appropriate weights to different associated nodes. Using the attention mechanism to assign higher weights to neighboring nodes or meta-path features that are more strongly associated with the link, the network computational efficiency can be improved when aggregating features at the node level and the association relationship level. In addition, it is also important to note in attribute embedding that the financial services on both sides of the link are of different types, and they have different attribute information, which indicates different meanings respectively. Since the formation mechanism of different types of nodes is different, there are correlations and influences between them, thus playing different effects on link prediction. If they are treated in the same way, a lot of important information will be lost, so they should be treated separately. After selecting the associated nodes, two levels of attention network are designed to calculate the attribute embedding values of the associated nodes, and separate processing is implemented.

Node level based attention mechanism is used to learn the importance of different nodes in the same set of associated nodes. Since the more similar the nodes are to each other, the stronger the association between them. So, in node level attention mechanism node similarity is taken as the probability of attention allocation and dot product is utilized to calculate the similarity of nodes. The reason for setting the attention level here is to assign higher weight to the associated nodes that are more similar to the destination node. As an example, to predict whether there is a link between Mashup service *Mi* and API service *Aj*, for an associated node of API service type *Al*, its probability of attention allocation in the set of association relations of the same class of *Aj* is:

$$\beta_{jl}^{\text{Node}} = \text{Similarity}\left(W_j, W_l\right) = W_j \cdot W_l \tag{3}$$

where *Wj* and *Wl* denote the embedded values of the attribute values of *Aj* and *Aj* and *AAl*, respectively, after the One-Hot encoding. Associated nodes of the Mashup type are in the same method of computed. The weight of *Al* in *Aj ejl Node* is computed by the Softmax function to calculate:

$$e_{jl}^{\text{Node}} = \text{Softmax}\left(\beta_{jl}^{\text{Node}}\right) = \frac{\exp\left(\beta_{jl}^{\text{Node}}\right)}{\sum_{m \in V_{ij}^{\gamma}} \exp\left(\beta_{jm}^{\text{Node}}\right)} \tag{4}$$

Represent the association relationship as $\gamma, \gamma \in$ {Pre, DCom, HCom, Par, Fun}, with the service node pair M_i, A_j, The set of nodes that are related to the service node pair V_{ij}^{γ}, embedded values of the association relationship γ

$$\delta_{ij}^{\gamma} = \sum\nolimits_{l \in V_{ij}^{\gamma}} \sigma\left(e_{jl}^{\text{Node}} \cdot W_l\right) \tag{5}$$

where σ denotes the activation function, here we use the hyperbolic sine function sinh and hyperbolic cosine function cosh ratio, i.e., the hyperbolic tangent function tanh which is the hyperbolic tangent function.

Attention at the association relation level is learned through a multilayer perceptron network that learns the probability of attention allocation for each type of association relation. Therefore, the association relation γ The attention allocation probability of an association relation is

$$\beta_{\gamma}^{Ass} = MLP\left(\delta_{ij}^{\gamma}\right) \tag{6}$$

Due to the difference in the number of association relationships between Mashup and API types, there is only one association relationship for Mashup type, and the weight of the Fun relationship is e_{Fun}^{Ass} is

$$e_{Fun}^{Ass} = \frac{\exp\left(\beta_{Fuss}^{Ass}\right)}{\exp\left(\beta_{Fun}^{ASs}\right) + \exp(W_i)} \tag{7}$$

Mashup Service M_i weights are:

$$e_i^{AsS} = \frac{\exp(W_i)}{\exp\left(\beta_{Fun}^{ASS}\right) + \exp(W_i)} \tag{8}$$

API type correlation γ weights e_{γ}^{AsS} are:

$$e_{\gamma}^{ASS} = \frac{\exp\left(\beta_{\gamma}^{Ass}\right)}{\sum_{\mu\in\{Pre,DCom,HCom,Par\}} \exp\left(\beta_{\mu}^{Ass}\right) + \exp\left(W_j\right)} \tag{9}$$

API Service A_j weights are.

$$e_j^{\mathrm{Ass}} = \frac{\exp\left(W_j\right)}{\sum_{\mu\in\{Pre,DCom,HCom,Par\}} \exp\left(\beta\beta_{\mu}^{ASs}\right) + \exp\left(W_j\right)} \tag{10}$$

The final attribute embedding value is the output of the association-level attention network, denoted as Z_i^{Attr} and Z_j^{Attr}:

$$Z_i^{Attr} = e_{Fun}^{Ass} \cdot \delta_{i,j}^{Fun} + e_i^{Ass} \cdot W_i \tag{11}$$

$$Z_j^{Attr} = \sum_{\gamma\in\{\mathrm{Pre},DCom,HCom,Par\}} e_{\gamma}^{Ass} \cdot \delta_{i,j}^{\gamma} + e_j^{Ass} \cdot W_j \tag{12}$$

Algorithm 1: Attribute Input Algorithm Based on Graph Attention Networks

---

Input: Financial services network $SNet = \langle V, E\rangle$;
API Node Properties $\{W_i, \forall i \in V^{API}\}$;
Mashup Node Properties $\left\{W_j, \forall j \in V^{\mathrm{Mashup}}\right\}$;
Associative Relationships { Pre, DCom, HCom, Par, Fun };

Output. API Property Embedded Value $\{Z_i^{Attr}, \forall i \in V^{API}\}$;
Mashup property embedding value $\left\{Z_j^{\mathrm{Attr}}, \forall j \in V^{\mathrm{Mashup}}\right\}$;
1: for $\gamma \in$ { Pre, DCom, HCom, Par, Fun } do
2: for $i \in V^{API}, j \in V^{\mathrm{Mashup}}$ do
3: Find neighbors based on association relationship V_{ij}^{γ} ;
4: for $k \in V_{ij}^{\gamma}$ do
5: if $k \in V^{API}$ then
6: Calculate the attention allocation probability of node k, eik^{Node} ;
7: else
8: Calculate the attention allocation probability of node k, ejk^{Node} ;
9: end if
10: end for
11: Calculating nodes i in Association Relationships γ embedded value δ_{ij}^{γ};
12: end for
13: Calculating Association Relationships i and nodes j the attention allocation probability β_{γ}^{ASS} ;
14: Calculating the final i and j attribute embedding values Z_i^{Attr} and Z_j^{Attr} ;
15: end for
16: return $\{Z_i^{Attr}, \forall i \in V^{API}\}, \left\{Z_j^{Attr}, \forall j \in V^{\mathrm{Mashup}}\right\}$;

---

4 Modeling Risk Propagation in Complex Financial Networks

To facilitate the description, similar to the study of information propagation models in networks, the process of risk propagation of service nodes in a complex financial ecosystem network, which may be presented in three states, risky, and repaired will be borrowed from some concepts in infectious diseases, and will be denoted by susceptible (Susceptible) services to denote nodes that have not yet appeared at risk, Infected services to denote nodes that have appeared at risk, and Recovered services to denote problematic nodes that have been repaired. The classical SIR (Susceptible- Infected- Recovered) model is used to describe the state change of each node, as shown in Fig. 2, in which each node is initially in the susceptible state (*Sus.*), at each moment t In this model, each node is initially susceptible to infection (inf), and at each moment they have $v(i, t)$ probability of being infected by a neighboring node, and for infected services, there is a $q(i)$ probability of repair.

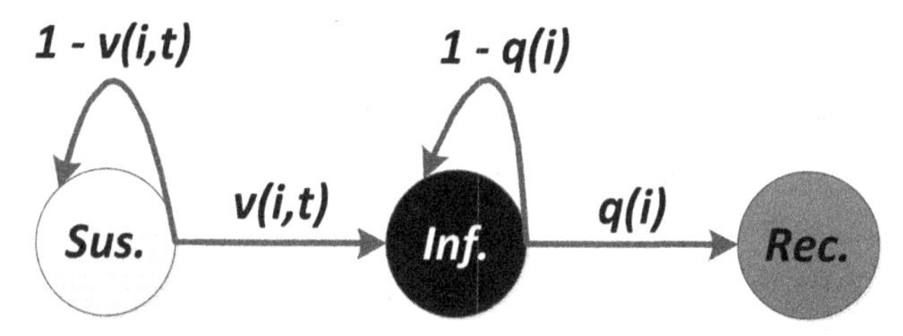

Fig. 2. Relationship of state transformation after the occurrence of risk at each node

Since the complex financial ecosystem network is a dynamic temporal network and each neighboring financial service node may be different at different moments,

$$v(i, t) = 1 - \Pi j \in N(i, t)[1 - \eta ji\, PI(j, t-1)] \tag{13}$$

Among other things. $N(i, t)$ For services i at the t the set of neighboring nodes at the moment, and ηji denotes the set of neighbor nodes of the service j. The probability that the occurrence of a problem will lead to the service i probability of having a problem, and $PI(j, t-1)$ denotes the probability that the service j has a probability of having a problem in $t-1$ the probability of being in the infected state.

This yields the probability that all services are in three different states, susceptible, infected and repaired, at each moment in time:

$$PS(i, t) = [1 - v(i, t)]PS(i, t-1) \tag{14}$$

$$PI(i, t) = v(i, t)PS(i, t-1) + [1 - q(i)]PI(i, t-1) \tag{15}$$

$$PR(i, t) = PR(i, t-1) + q(i)PI(i, t-1) \tag{16}$$

Based on the aforementioned characteristics of the spatial structure of complex financial systems, firstly, the risk propagation range is given *rFault* of the approximate *SCNet* calculation formula is as follow:

$$r_{SCNet}^{Fault} = \frac{1}{|V|}\sum\nolimits_{v\in V}\left[\frac{p_v^{Fault}}{\sum_{u\in V} p_u^{Fault}/|V|} \times r_v^{Fault}\right] = \frac{1}{N_{Sim}}\sum\nolimits_{i=1}^{N_{Sim}}\left[\frac{p_{v_i^{Src}}^{Fault}}{\sum_{u\in V} p_u^{Fault}/|V|} \times r_i^{Fault}\right] \quad (17)$$

where r_v^{Fault} denotes the node v is the risk propagation range when it is a risk source, and $\frac{p_v^{Fault}}{\sum_{u\in V} p_u^{Fault}/|V|}$ denotes the financial service node v the higher probability of occurrence of risk, the higher its contribution to the overall network risk propagation range.

Based on the above analysis, the following algorithm is given to calculate the range of network risk propagation in complex financial ecosystems:

Algorithm 2: Calculating the extent of network risk propagation in complex financial ecosystem

CalFaultSpreadRange($V_{CS}, V_{FM}, V_{MS}, V_{RP}, E_{FM\to CS}, E_{MS\to FM}, E_{RP\to MS}, E_{FM\to FM}, E_{MS\to MS}$)

$r_{SCN}^{Fault} = 0$;

$\overline{p_{Nd}^{Fault}} =$ **CalNdMeanFaultRate**($V_{CS}, V_{FM}, V_{MS}, V_{RP}$);

$N_{Sim} = |V_{CS} \cup V_{FM} \cup V_{MS} \cup V_{RP}| \times 10$;

$T_{TakeMeasures} = 100$;

For $i = 0; i < N_{Sim}; i = i + 1$

$v_i^{Src} = Rand(V_{FM} \cup V_{MS} \cup V_{RP})$;

$r_i^{Fault} =$ **SimFaultSpread**($v_i^{Src}, G_{SC}, T_{TakeMeasures}$);

$r_{SCNet}^{Fault} = r_{SCNet}^{Fault} + \frac{p_{v_i^{Src}}^{Fault}}{\overline{p_{Nd}^{Fault}}} \times r_i^{Fault}$;

End

$r_{SCNet}^{Fault} = \frac{r_{SCNet}^{Fault}}{N_{Sim}}$;

Return r_{SCNet}^{Fault};

End

5 Traceability Accuracy Verification of High Heat Node Setting Traceability Monitoring Node

In order to show that traceability monitoring nodes based on heat settings can effectively improve the accuracy of traceability, we give the formula for the difficulty of network risk traceability in complex financial ecosystems.

First, the difficulty of cyber risk traceability is given $d_{\mathrm{SCNet}}^{\mathrm{FaultTracing}}$ of the approximate calculation formula is as follows: where V_i^{Fault} denotes the first i set of financial service nodes at risk at the final moment of the second simulation, and $\overline{r_l^{\mathrm{TrueSrC}}}$ denotes the number of nodes in the first i average of the ranked estimates of the real risk sources calculated by several classical traceability algorithms, the less difficult it is to trace the network risk, and we use the three classical traceability algorithms of Rumor Center, Jordan Center, and DMP to calculate the estimation of each risk node (the node is the estimation of the probability of the risk source).

Algorithm3: The Difficulty of Tracing Network Risks in Complex Financial Ecosystems

CalFaultTracingDifficulty$(V_{CS}, V_{FM}, V_{MS}, V_{RP}, E_{FM\to CS}, E_{MS\to FM}, E_{RP\to MS}, E_{FM\to FM}, E_{MS\to MS})$

$d_{SCNet}^{FaultTracing} = 0;$

For $i = 0; i < N_{Sim}; i = i + 1$

$r_{i,RumorCenter}^{TrueSrc} =$ **FaultTracingByRumorCenter**$(v_i^{Src}, V_i^{Fault}, G_{SC});$

$r_{i,JordanCenter}^{TrueSrc} =$ **FaultTracingByJordanCenter**$(v_i^{Src}, V_i^{Fault}, G_{SC});$

$r_{i,DMP}^{TrueSrc} =$ **FaultTracingByDMP**$(v_i^{Src}, V_i^{Fault}, G_{SC});$

$$\overline{r_i^{TrueSrc}} = \frac{r_{i,RumorCenter}^{TrueSrc} + r_{i,JordanCenter}^{TrueSrc} + r_{i,DMP}^{TrueSrc}}{3};$$

$$d_{SCNet}^{FaultTracing} = d_{SCNet}^{FaultTracing} + \frac{\overline{r_i^{TrueSrc}}}{|V_i^{Fault}|};$$

End

$$d_{SCNet}^{FaultTracing} = \frac{d_{SCNet}^{FaultTracing}}{N_{Sim}};$$

Return $d_{SCNet}^{FaultTracing};$

End

6 Experiments and Analysis of Results

6.1 Dataset

In order to construct a complex financial service network, the data used in the experiments are crawled from real OPENAPI platforms ProgrammableWeb, HKSTP OPENAPI, and PingAn.com Web, including textual descriptions and attribute information of individual financial service nodes, as well as the interaction relationships between various types of objects contained in the financial service ecosystem. The data constructs a financial service network with a total of 48,044 nodes and 143,351 edges.

6.2 Simulation Calculation of Risk Propagation in Complex Financial Networks

Based on the aforementioned experimental dataset in March 2020, using Algorithm 1 to set the degree, meso-centricity, and tight-centricity of complex financial service network nodes as the baseline for risk node propagation sources, respectively, the propagation range shown in Fig. 3 can be obtained. From the figure, we can see that these center eigenvalues are large risk node propagation range is wider, once the risk brought about

Fig. 3. HKSTP OpenAPI, PingAn OpenApi, and ProgrammableWeb Examples

by the loss as well as the risk is greater. These eigenvalues are more important network topology features in the network, and changes in the eigenvalues will lead to changes in the propagation range of risk nodes, so analyzing and predicting the structure of the complex financial services network through the above eigenvalues can positively affect the risk propagation of the complex financial ecosystem network as well as traceability. Figure 4 Effect of degree, betweenness centrality, tightness centrality, and pagerank of financial service nodes on the extent of risk propagation.

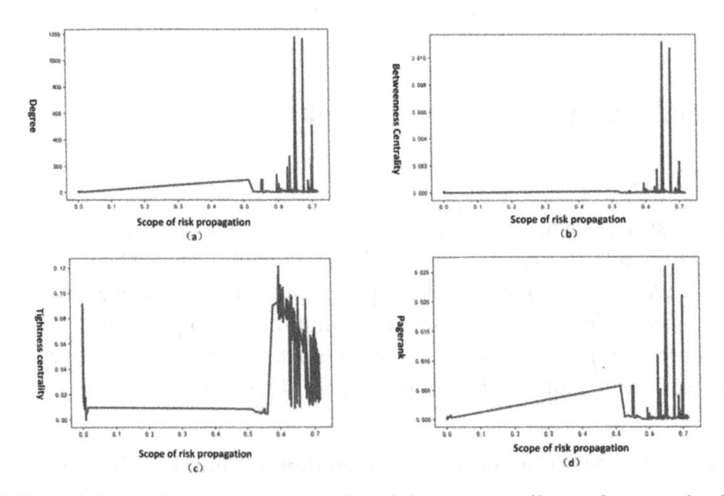

Fig. 4. Effect of degree, betweenness centrality, tightness centrality, and pagerank of financial service nodes on the extent of risk propagation

6.3 Traceability Accuracy Verification

After completing the experiments of this experimental group according to Algorithm 4, we get the experimental results as shown in Fig. 4 and Fig. 5. It can be seen that when fault propagation occurs in the complex financial ecosystem, based on the top ranked hotness predicted by the model of this paper 10% of API states for traceability, the traceability difficulty is the lowest, which is closer to the difficulty of traceability based on the states of all nodes in the network of complex financial ecosystems, and the proportion of faults found by traceability monitoring nodes based on heat settings is high. Figure 5(a)–(f) shows the distribution of traceability difficulty in 30 trials when different measures (Deg, BC, CC, Pagerank and Pop) are used as the basis for choosing the deployment location of the traceability monitoring agent in the complex financial ecosystem network, and Fig. 6 shows the average value of the traceability difficulty of different measures. The average value, it can be clearly seen that the traceability difficulty is lower when deploying traceability monitoring nodes based on node hotness. Therefore, using the node heat degree predicted by the model in this paper as the basis for choosing the deployment location of the monitoring agent (which monitors the propagation of faults as input information for the traceability algorithm) in the service ecosystem can

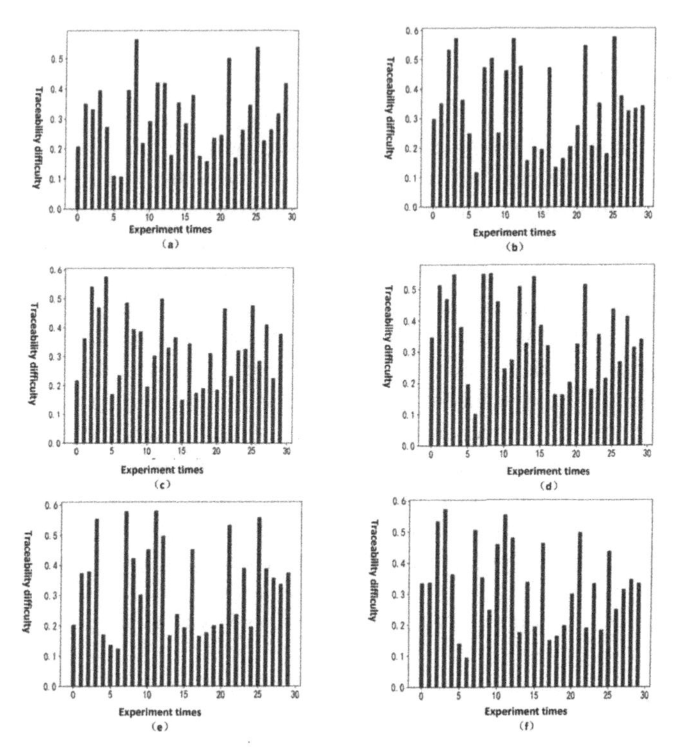

Fig. 5. Distribution of traceability difficulty over 30 experiments when the choice of monitoring agent deployment location in a complex financial ecosystem is based on different experiment times.

effectively reduce the difficulty of fault traceability, improve the accuracy of traceability, and safeguard the security and accountability of the service ecosystem.

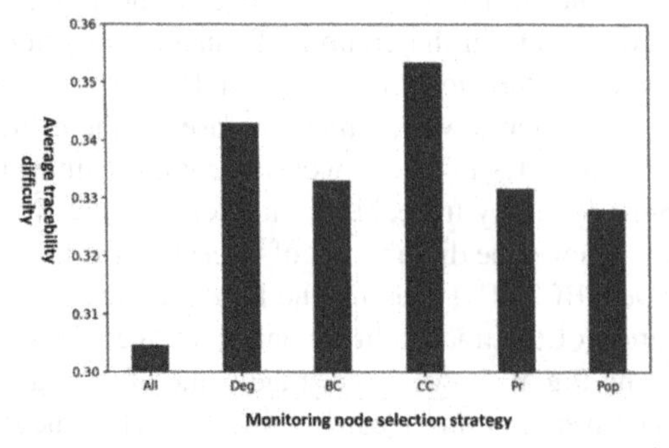

Fig. 6. Average values of traceability difficulty when different measures (degree, betweenness centrality, tightness centrality, Pagerank, and heat as predicted by the model in this paper) are used as the basis for choosing the deployment location of the monitoring agent in complex financial ecosystems

7 Conclusion

This research focuses on the application prospects of new artificial intelligence methods in risk control in the financial industry, and captures the reality that the business of each branch financial institution under a large financial control group is interdependent and unable to carry out effective risk prevention, control and accountability, and gives the method of constructing a complex financial ecosystem based on the attention network of heterogeneous graphs. The method can accurately portray the attributes and interrelationships of each financial entity in the financial ecosystem, and predict its spatial structure evolution based on historical eigenvalues, so as to portray the network structure of the entire financial ecosystem and lay a theoretical foundation for the deployment of risk prevention, control and accountability mechanisms. Meanwhile, the financial risk propagation model and its corresponding traceability method are proposed, which can effectively analyze the risk propagation range of financial entity nodes after the emergence of risk, and can block the risk from the source, improve the risk resistance and stability of the whole financial ecosystem, and enhance the risk disposal ability.

Funding. This work was supported by China Postdoctoral Science Foundation (2022M712982).

References

1. Boccaletti, S., et al.: The structure and dynamics of multilayer networks. Phys. Rep. **544**(1), 1–122 (2014)
2. Chen, Y.Z., Huang, Z.G., Zhang, H.F., Eisenberg, D., Seager, T.P., Lai, Y.C.: Extreme events in multilayer, interdependent complex networks and control. Sci. Rep. **5**, 17277 (2015)

3. Cheng, X., Wu, J., Liao, S.S.: A study of contagion in the financial system from the perspective of network analytics. Neurocomputing **264**(Suppl. C), 42–49 (2017)
4. Dehmamy, N., Buldyrev, S.V., Havlin, S., Stanley, H.E., Vodenska, I.: A systemic stress test model in bank-asset networks. arXiv e-prints (2014)
5. Delpini, D., Battiston, S., Riccaboni, M., Gabbi, G., Pammolli, F., Caldarelli, G.: Evolution of controllability in interbank networks. Sci. Rep. **3**(15), 1626 (2013)
6. Fang, M., Shi, P., Shang, W., et al.: Locating the source of asynchronous diffusion process in online social networks. IEEE Access **6**, 17699–17710 (2018)
7. Alexandru, R., Dragotti, P.L.: Rumour source detection in social networks using partial observations. In: IEEE Global Conference on Signal and Information Processing, pp. 730–734 (2018)
8. Hu, Z., Shen, Z., Tang, C., et al.: Localization of diffusion sources in complex networks with sparse observations. Phys. Lett. A **382**(14), 931–937 (2018)
9. Hu, Z., Shen, Z., Han, J., et al.: Localization of diffusion sources in complex networks: a maximum-largest method. Physica A **527**, 121262 (2019)
10. Fletcher, K.: Regularizing matrix factorization with implicit user preference embeddings for web API recommendation. In: 2019 IEEE International Conference on Services Computing (SCC). IEEE (2019)
11. Cao, B., Liu, X.F., Rahman, M.M., et al.: Integrated content and network-based service clustering and web APIs recommendation for mashup development. IEEE Trans. Serv. Comput. **13**(1), 99–113 (2020)
12. Xie, F., Chen, L., Ye, Y., et al.: Factorization machine based service recommendation on heterogeneous information networks. In: 2018 IEEE International Conference on Web Services (ICWS). IEEE (2018)

Short Paper Track

Accelerate SVM Training with OHD-SVM on GPU

Chao-Chin Wu(✉), De-Xang Wang, and Lien-Fu Lai

Department of Computer Science and Information Engineering University, National Changhua University of Education, Changhua, Taiwan
ccwu@cc.ncue.edu.tw

Abstract. The SVM (Support Vector Machine) faces a huge challenge on reducing training time in the presence of large data sets due to its high computational cost. The SMO (Sequential Minimal Optimization) algorithm is used to solve the quadratic programming problem that arises during the training of SVM. OHD-SVM leverages the power of GPU (Graphics Processing Unit) to accelerate SVM training and it is the fastest GPU implementation so far. OHD-SVM adopts two-level heuristics to select SMO working set. In this paper, we propose a method that narrows the search domain for selection of the working pair in each SMO iteration. Six datasets from LibSVM are used for performance evaluation. Experimental results demonstrate that in most cases our method outperforms OHD-SVM with similar accuracy. At best, the training can be about seven times faster. For several cases, both the training time and the accuracy are improved. In the best case, the accuracy is improved about ten times with 1.46 speedup of training time.

Keywords: SVM · SMO algorithm · training time · GPU · parallel computing

1 Introduction

SVM (Support Vector Machine) is a popularly used machine learning method. Despite the vigorous development of artificial neural networks in recent years, SVM is still used in many fields and achieves quite excellent results. In the past three years, there have been more than 10,000 papers and works related to SVM in the IEEE electronic database.

The main problem with SVM is its long training time because it is computationally expensive, especially for large input datasets. The SVM training process is a quadratic programming problem and Sequential Minimal Optimization (SMO) is widely used for solving the problem in SVM [1]. In the presence of large-scale data sets nowadays, it is difficult to train an SVM model in a time-efficient and high-accuracy manner. To tackle these problems, various reduction algorithms have been introduced to prun the size of the training set while the error rate is kept as low as possible [2]. Another kind of solutions is to leverage the computing power of GPU (Graphics Processing Unit) to speed up the SVM training. Several GPU implementations of SVM training such as GPUSVM [3], ThunderSVM [4], and OHD-SVM [1] have been introduced, where OHD-SVM is the fastest. OHD-SVM achieved speed-up up to 12 times in comparison

S. Zhang et al. (Eds.): BigData 2023, LNCS 14203, pp. 209–217, 2023.
https://doi.org/10.1007/978-3-031-44725-9_15

with the second fastest GPU implementation. When compared with the sequential SVM, the performance of OHD-SVM can be improved even by nearly a hundred times.

OHD-SVM adopted the SMO-based algorithm to solve the quadratic programming problem. In the GPU, it is impossible to perform SMO on all training data like the original SVM because of the limited hardware resources. Only part of the training data, called the working set, is used during each SMO iteration. Working set selection in the SMO algorithm is a very important part of the SVM training algorithm, determining the training time and the final SVM model. In SMO, a working pair is selected using either first order or second order heuristic. The first order heuristic determines whether the solution is optimal or not. If not, second order heuristic is used to select a new working pair. The first point of the working pair is selected in the first heuristic and the second point is picked in the second heuristic. The penalty parameter C of the SVM model is used in the selection of the working set and the working pair in each iteration. The C value determines the search space.

In this paper, we use the C value for working set selection but $0.5C$ for working pair selection. In this way, the search space of working pair selection is reduced significantly. We have used six datasets from LibSVM [5] for performance comparison between our method and OHD-SVM on the NVIDIA A100 GPU. Experimental results showed our method outperforms OHD-SVM on training time with similar accuracy for most cases. In some cases, both the training time and the accuracy are improved. The best improvement on training time is eight times and accuracy ten times.

2 Related Work

SVM faces a huge challenge on reducing training time in the presence of large data sets due to its high computational cost. To reduce the training time, researchers proposed approaches based on the mathematical properties to solve the convex quadratic programming optimization problem efficiently. For instance, Active Set method is an effective method for solving quadratic programming problems with inequality constraints [6]. If we know the active constraints at the optimal point, we can remove all the inactive constraints. That is, transform some inequality constraints into equality constraints, reducing the complexity of searching for solutions. Note that the derivation of SVM happens to be a quadratic programming problem with inequality constraints. In the Active Set approach, it is necessary to judge which inequality constraints have no effect on the result, and which inequality constraints can be turned into equality constraints. For general problems, if we turn all inequality constraints into equality constraints, then the feasible region may become an empty set. Therefore, the following strategy is adopted instead.

Define a working subset of constraints that includes a subset of equality constraints that need to ensure that the feasible domain is not empty. First try to find the optimal point under the feasible domain corresponding to the current working subset. If this point does not satisfy another constraint, we turn that constraint into an equality constraint and include it in the working subset. Then, do it again until all constraints are satisfied. However, because the feasible domain of equality constraints is smaller than that of inequality constraints, the optimal point obtained from the feasible domain corresponding to the current working subset is not necessarily the optimal value of the original problem. Each working subset corresponds to a linear equation system and also corresponds to an extreme value. The main purpose of each iteration is to update the working subset until an extreme point is found. The conditions that the most value points need to meet are as follows.

- Satisfy the KKT conditions of all current equality constraints.
- After the original inequality constraints in the working set become equality constraints, the corresponding $\lambda \geq 0$.
- The obtained result should also satisfy other constraints that are not in the working set.

Since SVM is widely used, various SVM tools are continuously proposed. LibSVM proposed by Chang and Lin [5] is a Library of SVM, written in C++. This software has been actively developed since 2000 with the goal of helping users easily implement SVMs into their applications. LIBSVM has gained widespread popularity in machine learning and many other fields. To address the problem of long SVM training time, several GPU implementations have been proposed. GPUSVM [3] proposed by Catanzaro et al. is an open source code and uses CUDA to implement the SMO algorithm. It supports dense data, linear binary SVM classification (C-SVM), multinomial, Gaussian, and sigmoid kernels. Both first and second order differentiation are used to select the SMO working set. Next, use second order differentiation to reduce the number of training iterations required for convergence. GPUSVM has better performance than LIBSVM on dense datasets and is one of the best GPU SVMs ever released. Herrero-Lopez introduced a GPU multi-class SVM training in [7]. This is the first GPU SVM implementation that allows multi-class classification in a one-vs-all fashion except for binary classification problems. Zeyi et al. proposed ThunderSVM [4], an open source SVM software toolkit that utilizes GPU and multi-core CPU to achieve high performance. ThunderSVM supports all functions of LibSVM, including classification, regression and one-class SVM, and uses the same command instructions, so that LibSVM users can easily apply it. Experiments showed that ThunderSVM is faster than LibSVM. The OHD-SVM [1] GPU implementation proposed by Jan Vaněk et al. in 2017 proposed an SVM training algorithm with a two-layer optimization architecture. The core idea of its two-layer optimization architecture is to optimize the original SVM training process, reduce it to local optimization, and reduce large problems to small problems for optimization. Improve the overall performance without affecting the accuracy rate. Next, for the kernel value problem, a software cache system is proposed to store the calculated kernel value in the cache to avoid subsequent repeated calculations. To the best of our knowledge, so far OHD-SVM is the fastest GPU implementation of SVM training in general.

3 Proposed Method

The SVM model has two very important parameters: C and gamma. Among them, C is the penalty coefficient, used to control the tolerance of the systematic outliers, that is, the tolerance for errors. The higher the C, the less tolerance for errors and easy overfitting. If C approaches to infinity, then the classifier will approach the usual maximum margin classifier that does not allow any errors. On the other hand, if the value of C is too small, the classifier will be too tolerant to classification errors, so the classification accuracy will be poor. In short, too large or too small,C will lead to poor model generalization ability.

As for the gamma, it is a parameter that comes with the function after the Gaussian RBF function is selected as the kernel. According to *scikit-learn*, intuitively, the gamma parameter defines how far the influence of a single training example reaches, with low values meaning 'far' and high values meaning 'close'. The gamma parameters can be seen as the inverse of the radius of influence of samples selected by the model as support vectors. The larger the gamma, the fewer support vectors, and the smaller the gamma value, the more support vectors. The number of support vectors affects the speed of training and prediction.

In the GPU, it is impossible to perform SMO on all vectors like the original SVM. Due to hardware limitations, only part of the vectors can be selected in the GPU in each iteration of the SMO algorithm. These selected vectors in each iteration are together called the working set that is very important for the model accuracy we can obtain. One pair of working sets in the original SMO can be selected by first-order differentiation or second-order differentiation. In the local method of OHD-SVM, the first order differential is used to judge whether the subproblem has been optimized. If not already optimized after that, use second-order differentiation to select a new pair of working sets. OHD-SVM requires to select N_{WS} points for the new working set in each global iteration, where N_{WS} is the size of working set. In the first order selection, take first $N_{WS}/2$ values instead of one point for a pair of α. Working set selection in OHD-SVM consists of three steps.

In Step 1, Use first order heuristic to select $N_{WS}/4$ points. All the vectors in the current working set are sorted based on their values of $y_k \cdot g_k$, where y_k represents the label y value corresponding to sample k; g_k represents the gradient value corresponding to sample k. For each sample k, its priority depends on the value of $y_k \cdot g_k$, The sets of $vector_{up}$, and $vector_{low}$ consist of the vectors with positive $y_k \cdot g_k$ and vectors with negative $y_k \cdot g_k$, respectively. Then select $N_{WS}/4$ vectors from $vector_{up}$ and $vector_{low}$, respectively, according to per vector priority.

In Step 2, select vectors for the current working set $current_{WS}$. Sequentially select $N_{WS}/2$ vectors from the following three sets based on the SMO outcome: (i) the set of free vectors ($0 < \alpha < C$), (ii) the set of lower vectors ($\alpha \leq 0$), (iii) the set of upper vectors ($\alpha \geq C$). As long as there are enough $N_{WS}/2$ vectors selected, stop picking.

Finally, in Step 3, the collection of these vectors selected in Steps 1 and 2 becomes a new working set WS^{new} of size N_{WS}, which will be used in the next SMO iteration.

OHD-SVM used SMO as the local solver used to optimize the sub-problem consisting of points in the current working set. The whole local solver is implemented in one CUDA kernel. Shared memory is used for parallel reductions when selecting the current working pair in each iteration and to exchange data between threads when updating α coefficients. The first order will select the first point in the working pair and the second order then select the second point in the working pair. After these two points are optimized against each other, update the coefficient values and the whole gradient vector. When the solution for the local sub-problem converges, update the gradient vector for all training points for the selection of the new working set.

When performing the selection in the first order heuristic, the following two conditions are used in OHD-SVM to select the first point in the working pair, where α_i represents the α coefficient of the first point i.

$$(y_i > 0 \text{ and } \alpha_i < \text{C}) \text{ or } (y_i < 0 \text{ and } \alpha_i > 0) \tag{1}$$

$$(y_i > 0 \text{ and } \alpha_\text{i} > 0) \text{ or } (y_i < 0 \text{ and } \alpha_\text{i} < \text{C}) \tag{2}$$

Following the concept of the Active Set method, this paper proposes modifying the inequality constraints in SMO by changing the original constraint inequality $0 < \alpha < C$ to $0 < \alpha < C/2$. That is, we modify the conditions (1) and (2) to the following ones for the selection of the first point in the working pair

$$(y_i > 0 \text{ and } \alpha_i < \text{C}/2) \text{ or } (y_i < 0 \text{ and } \alpha_i > 0) \tag{3}$$

$$(y_i > 0 \text{ and } > \alpha_i\, 0) \text{ or } (y_i < 0 \text{ and } \alpha_i < \text{C}/2) \tag{4}$$

Similarly, we replace the original C value with $C/2$ in the constraint condition for the selection of the second point. The proposed method refers to the concept of Active Set method to reduce the search domain.

4 Experimental Results

For all the tests we used a Server with Intel Xeon Gold 622R, 16-core CPU clocked at 2.9 GHz with 128 GB RAM at 2.9 GHz, and NVIDIA A100 GPU, 6912-core GPU clocked at 1410 MHz with 40 GB RAM. The precision format adopted in A100 is Tensor Float (TF32), the software operating system version is Ubuntu 20.04.1 LTS 64bits, and CUDA version 11.1. The detailed configurations of the CPU and the GPU are listed in Table 1. As for the datasets, they come from the LibSVM data page (https://www.csie.ntu.edu.tw/cjlin/libsvmtools/datasets/) and their characteristics are shown in Table 2.

First, we compare the training time and the results are shown in Fig. 1. The speedup is derived by the following equation.

$$speedup = T_{OHD\text{-}SVM} \,/\, T_{Ours}, \tag{5}$$

where $T_{OHD\text{-}SVM}$ and T_{Ours} represent the training times of OHD-SVM and ours, respectively. Nine combinations of the C and the gamma (G) values are tested to demonstrate the performance of our proposed method, where the C values are 1, 100 and 500 while the G values are 0.001, 001 and 0.05.

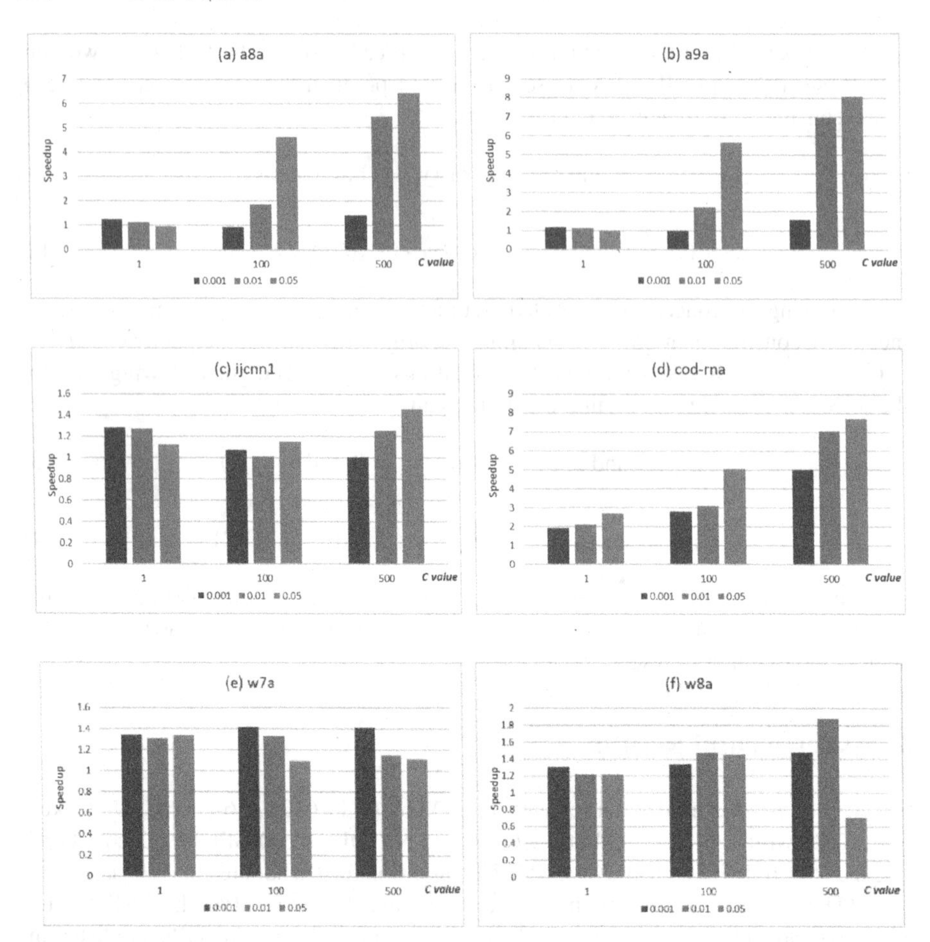

Fig. 1. Speedup comparison between different combinations of the values of G and C. The legend shows the G values.

In general, the best speedup is obtained when the C value is 500 except the dataset *w7a*. It can be expected because our method can obtain more advantage when C is increased, where C defines the squared search space in the SMO algorithm. Our method cuts the search space one fourth to find the working pairs. The best speedup is about eight times coming from *a9a*. As for the G value, the larger the G value, the fewer support vectors, and vice versa. The number of support vectors affects the speed of training. When the G value is increased, usually our method can provide better speedup especially for the larger C value.

Table 1. The configurations of CPU and GPU.

CPU Intel(R) Xeon(R) Gold 6226R		GPU NVIDIA A100	
Number of Cores	16	Number of GPUs	1
Number of Threads	32	Thread Processors	6912
Clock Speeds	2.9 GHz	Clock Speeds	1410 MHz
Memory Size	128 GB	Memory Size	40 GB
Memory Type	DDR4-2933	Memory Type	HBM2

Although the improvements of the training time for datasets *ijcnn1*, *w7a* and *w8a* are about between 1.2 to 1.4, our method can also improve their accuracy significantly for the first two datasets in many cases, as shown in Table 3. For instance, the accuracy of *ijcnn1* can be improved from 9.5 to 90.5. The accuracy of w8a obtained by OHD-SVM is higher than 95%, leaving a very small space of improvement. Nevertheless, our method has poor performance on accuracy for the dataset *cod-rnd* although the training time can be shortened. Compared with other datasets, the data dimension of *cod-rna* is much lower, even the original OHD-SVM is prone to poor convergence. More and further experiment and analysis are required in the future.

Table 2. List of Used Datasets from LibSVM.

Name	Training Size	Testing Size	Feature
a8a	22,696	9,865	123
a9a	32,561	16,281	123
ijcnn1	49,990	91,701	22
cod-rna	59,535	271,617	8
w7a	24,692	25,057	300
w8a	49,749	14,951	300

Table 3. Accuracy comparisons between OHD-SVM and ours. In the table, OHD represents the OHD-SVM method.

(G, C)	Method	a8a	a9a	ijcnn1	cod-rna	w7a	w8a
(0.001, 1)	OHD	82.82	59.87	9.5	89.89	2.95	97.24
	Ours	76.33	76.38	90.5	33.33	97.05	96.96
(0.001, 100)	OHD	76.33	23.62	9.5	62.59	2.95	96.96
	Ours	76.33	76.38	90.5	37.84	98.06	97.68
(0.001, 500)	OHD	23.67	76.38	9.5	62.59	2.95	98.75
	Ours	76.33	76.38	90.5	67.43	98.57	98.44
(0.01, 1)	OHD	31.39	84.87	12.62	46.58	98.34	98.15
	Ours	76.33	84.87	90.5	66.71	97.31	96.96
(0.01, 100)	OHD	84.94	84.99	90.5	92.57	98.8	99.09
	Ours	78.73	77.07	90.54	67.15	98.83	98.82
(0.01, 500)	OHD	84.45	84.6	9.5	91.75	98.73	99.25
	Ours	79.44	79.17	91.51	33.35	98.79	98.98
(0.05, 1)	OHD	85.14	85.09	20.45	90.52	98.57	98.86
	Ours	76.4	76.38	90.5	50.28	97.62	97
(0.05, 100)	OHD	82.69	82.94	90.08	92.3	98.67	99.43
	Ours	81.39	80.92	92.53	35.54	98.73	99.26
(0.05, 500)	OHD	80.46	81.02	9.5	90.61	98.54	99.45
	Ours	81.02	80.2	94.32	37.31	98.61	99.45

5 Concluding Remarks

In this paper, we proposed a method to improve OHD-SVM on the training time, where OHD-SVM is a fast GPU implementation of SVM training. We narrowed down the search space of the working pair selection in the SMO-based algorithm. The preliminary results demonstrate that our method has very outstanding performance. For most experimental cases, the training time is reduced significantly with little accuracy sacrificed. In some cases, the accuracy can be also enhanced dramatically. Furthermore, several important parameters of SVM need to be selected before we perform the training. Various heuristics have been proposed to determine the best settings of the parameters. Our method can be also adopted to accelerate the search of the optimal combination of the SVM parameters. More study and analysis on our proposed will be conducted in the future.

References

1. Vanek, J., Michalek, J., Psutka, J.: A GPU-architecture optimized hierarchical decomposition algorithm for support vector machine training. IEEE Trans. Parallel Distrib. Syst. **28**(12), 3330–3343 (2017)
2. Ghaffari, H.R.: Speeding up the testing and training time for the support vector machines with minimal effect on the performance. J. Supercomput. **77**(10), 11390–11409 (2021). https://doi.org/10.1007/s11227-021-03729-0
3. Catanzaro, B., Sundaram, N., Keutzer, K.: Fast support vector machine training and classification on graphics processors. In: Proceedings of the 25th International Conference on Machine Learning, pp. 104–111 (2008)
4. Wen, Z., Shi, J., Li, Q., He, B., Chen, J.: ThunderSVM: a fast SVM library on GPUs and CPUs. J. Mach. Learn. Res. **19**(1), 797–801 (2018)
5. Chang, C.-C., Lin, C.-J.: LIBSVM: a library for support vector machines. ACM Trans. Intell. Syst. Technol. **2**(3), 1–27 (2011)
6. Scheinberg, K.: An efficient implementation of an active set method for SVMs. J. Mach. Learn. Res. **7**, 2237–2257 (2006)
7. Herrero-Lopez, S., Williams, J.R., Sanchez, A.: Parallel multiclass classification using SVMs on GPUs. In: Proceedings of the 3rd Workshop on General-Purpose Computation on Graphics Processing Units, pp. 2–11 (2020)

Construction of Data Security System

Bingbing Yu[1(✉)] and Jiefan Hu[2]

[1] Chief Information Security Officer, Financial Industry, Beijing, China
margie2002@sina.com

[2] Data Governance Senior Manager, Financial Industry, Shanghai, China

Abstract. With the progress of science and technology, the development of new technologies, the in-depth application of big data, artificial intelligence, and cloud computing, data has gradually been transformed from information assets to production factors. Data breach, abuse, tampering and other security issues will cause great harm to enterprises, and even affect national security, societal order, public interests and market stability. Therefore, on the basis of meeting the fundamental business needs of enterprises, carrying out business and daily operation management, promoting the application and sharing of data, mining and realizing data value, strengthening data protection capabilities, and ensuring the safe flow of data are also the key points of data management work.

Keywords: Data Security · Data Classification · Data Lifecycle

1 Introduction

Information technology is constantly developing, and the basic business, core processes, inter industry transactions and activities of enterprises have all been run on information support carriers. The information generated by production and operation is gradually transformed into digital assets in different forms and circulated in information systems. With the progressing of science, the development of new technologies, and the in-depth implementation of big data, artificial intelligence, and cloud computing, data has gradually evolved from an information asset to a factor of production. Data breach, abuse, tampering and other security issues will cause significant harm to the enterprise, and even have an impact on national security, societal order, public interests and market stability. Therefore, on top of meeting the basic business needs of an enterprise, carrying out the business, and managing the day-to-day operation, it is important to incorporate promoting the utilization and sharing of data, mining and realizing data value, strengthening data protection capabilities, and ensuring the safe flow of data into the data management work.

In IT domain, the meaning of the word "security" has changed, from the security of the carriers of information systems to the security of data, from the technical security to the effective protection of and the legitimate utilization of data. Data security refers to taking necessary measures to ensure that data is effectively protected and legitimately utilized, as well as having the ability to ensure the "security" continuously.

S. Zhang et al. (Eds.): BigData 2023, LNCS 14203, pp. 218–229, 2023.
https://doi.org/10.1007/978-3-031-44725-9_16

2 Data Security Framework Based on Data Lifecycle

The data lifecycle includes data collection, data transmission, data storage, data usage, data deletion, and data destruction.

Data Security focuses on protecting data against unauthorized access and corruption during the whole life cycle of data. It covers a set of relevant standards, technologies, frameworks and processes.

Data Security includes the planning, development, and execution of data security policies and procedures, which provide proper authentication, authorization, access, and audit on data and information assets [1].

So, enterprises should develop a data security framework based on the security of the entire data lifecycle. This framework consists of three parts: security management, process and mechanism, and technologies and tools. Below is an example of data security framework (Fig. 1).

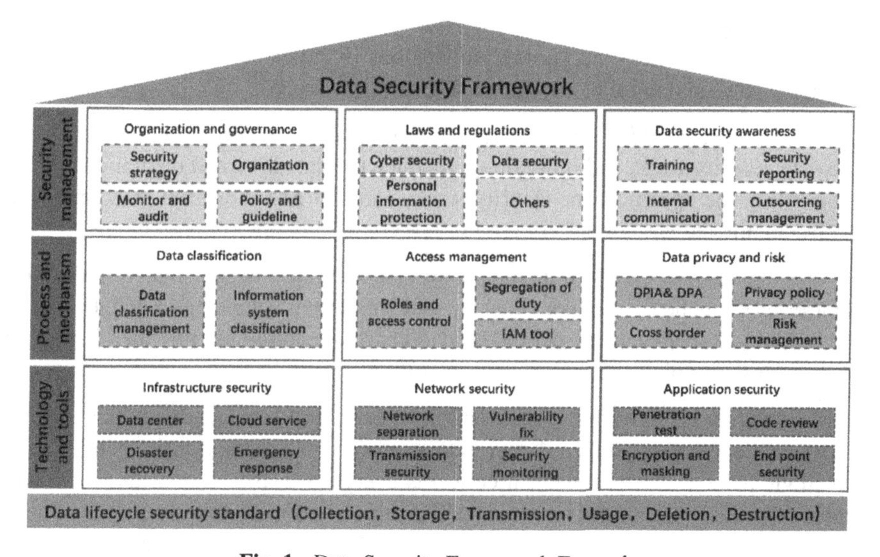

Fig. 1. Data Security Framework Example

2.1 Security Management

Principles

To ensure data security, the following principles shall be followed when processing data:

Legality and Compliance: It is necessary to ensure the data activities are legitimate and compliant to regulations throughout the entire data lifecycle.

Clear Purpose: A data security protection strategy should be developed to clarify the security protection goals and requirements for each stage of the data lifecycle.

Controllable Whole Process. Data security control mechanism and technical measures that match the security level of data should be adopted to ensure the confidentiality, integrity, and availability of data at all stages of the its lifecycle, and to avoid unauthorized access, destruction, tampering, leakage, or loss of data throughout the entire data lifecycle.

Dynamic Control. The security control strategy and security protection measures of data should not be one-off or static, but be able to adjust in real-time and dynamically to the factors such as business requirements, security environment attributes, and behaviors of system users.

Consistency of Rights and Responsibilities. The relevant departments and their responsibilities for data security protection should be clearly defined in organizations. The departments and its staff should actively implement the required measures and take their responsibilities of data security protection [2].

Security Management

The security management incorporate dimensions like data security organization and governance, laws and regulations, and data security awareness.

Data Security Organization and Governance. In addition, data compliance is also part of the data security work, and it important to abide the data laws and regulations. When implementing the specific data security work, national and industrial standards have provided significant guidance.

Laws and Regulation. In addition, data compliance has also been included in the work of data security, so the three major data laws and regulations should be abided. In order to better implement specific work, national and industry standards have provided significant guidance for the implementation of data security work.

Security Awareness. With the promulgation of laws and regulations, enhancing data security awareness is crucial for better data security work. Data security is not the job for a single person or department, it concerns everyone in the organization who have been involved in business and operations. The improvement of data security awareness comes from not only the training and communication to internal employees, but also the management of data security awareness of the third parties and the outsourcing counter parties. At the same time, regular audits and monitoring audits should be conducted on routine works to ensure that data security work is effectively implemented according to the designed security plan and process.

2.2 Process and Mechanism

The second layer of the data security framework, process and mechanism, includes data classification, access management, and privacy protection.

Data Classification

Data classification is the first step in data security governance. The hierarchical classification of data covers not only structured data, but also unstructured data. It includes data stored in information systems and data existing in business processes and business documents as well.

Data Access
The core of data security is the management of data access. Accesses should be managed following the principles of minimum fit-for-purpose and role-based access control.

Privacy Protection
Privacy protection is an important part of data security. Except for meeting the internal data security requirements, it is also necessary to meet all the requirements under applicable laws and regulations.

2.3 Technology and Tools

Data security cannot be separated from the security of data carrier, it requires the careful design from infrastructure security, to network security, then to application system security. There are a number of technologies and tools for data security, such as encryption, identity authentication and access control, data leakage prevention tools, monitoring logs, etc.

3 Data Security Work Practice Based on Risk Priority

3.1 Data Security Practice Steps

After determining the data security framework, the enterprise need to put the data security work into practice. It is known to all that data security is a seamless and endless task, and like a circle, it has no starting or ending point. One can never be never secured enough, but can always be more secured. Back to reality, it is not feasible to invest in security endlessly. So, we need to put the data security work into practice based on the risk priorities.

Although security is a common requirement of enterprises, the risk faced by each enterprise is different due to their natures of business and operations, and the subjective regulations.

In order to find a balance between security and cost, and to more accurately implement the data security work, the following four-step approach can be adopted to determine the direction and projects of data security. Literally, the four steps are data security requirements collection, data security gap analysis, data security improvement plan development, and rigorous implementation of security projects (Fig. 2).

Needs Collection
When collecting data security requirements, it is necessary to consider meeting the requirements of laws and regulations, national standards, and industry standards, as well as meeting the needs of business development to reduce the risks faced by data usage.

Gap Analysis
Based on the requirements and the understanding of existing data security measures of

Fig. 2. Data Security Work Practice Based on Risk Priority

the enterprise, analyze the usage of data in business processes and identify the gap with the requirements.

Action Plan
Based on the gap and the actual enterprise operation situation, considering the critical data security needs, a focused and feasible data security improvement plan shall be worked out.

Implementation
Finally, based on the established data security improvement plan, implement the project rigorously, check the project progress regularly, and ensure that project activities are completed in accordance to the plan.

3.2 Content of Data Security Practice

Once the data security risks are confirmed and the direction of work is determined for the enterprise, the data security work should be implemented. The following are several important tasks.

Data Classification
The core of data security lies in the management of data security throughout the entire data lifecycle. Implementing data lifecycle security management can further clarify the data protection requirements for each stage of the data lifecycle. It helps to allocate data protection resources and costs reasonably and establish a comprehensive data lifecycle protection mechanism.

In order to reasonably allocate data protection resources and costs, it is necessary to implement data classification. This will further clarify the targets of data protection and help enterprises to allocate theta protection resources and costs reasonably and to implement data security work in a focused manner, and lead to satisfactory returns. At the same time, a standardized data classification management system can enhance the safety of data sharing between organizations or industries, which is conducive to mining and realizing data value.

Data classification is the first step of data security work. The hierarchical classification of data covers not only the structured data, but also the unstructured data, which includes the data stored in information systems, and the data exists in business workflows and documents as well.

Data classification can be developed according to integrity, confidentiality, and availability. It is also a good practice to classify data based on its confidentiality level, like basic protection, special protection, and high protection. With the data classification, the corresponding protection measures should be set accordingly. The following are example (Table 1) of data classification and the corresponding data security requirements.

Table 1. Examples of Data Security Classification and Protection Measures

Confidential level	Data security requirements
Public	There are no specific IT security requirements
Internal	User authentication System accesses should be managed with permission control tools The privileged access of data containing IT assets must be managed with privileged access management tools The information system must be connected to the security monitoring tool before carrying out any day-to-day user activity The information system must be connected to and maintained in the vulnerability management tool Confidentiality labels must be in place There must be the function to prevent data loss and there-fore restricts the data flows outside the organization Data sharing with anybody outside of the organization should be limited and it must be explicitly approved by the data owner
Restricted	On top of the above User authentication based on multi factor authentication The information system must be connected to the security monitoring tool before conducting any detailed user activity There must be the function to protect data loss and there-fore restricts the data flows inside and outside the organization It is necessary to encrypt the static data inside the organization environment and the data in transit (external/internal) Data can only be shared with a limited number of staff who have signed the Non-disclosure agreements and a limited number of vendors. The data sharing should be limited to the data owner or their designated personnel

(*continued*)

Table 1. (*continued*)

Confidential level	Data security requirements
Confidential	On top of the above Clearly manage the approvals of the access granted The information system must be connected to security tools to obtain complete user activity and security logs Data sharing is strictly limited to the data owner or personnel approved by the data owner

Data Encryption

Data encryption is a long-history technology that converts plaintext into cipher-text through encryption algorithms and encryption keys, while decryption involves restoring ciphertext to plaintext through decryption algorithms and decryption keys. The core of it is Cryptology. Data encryption is still the most reliable way for computer systems to protect information. It uses cipher graph to encrypt information, to achieve information concealment, and thus to protect information security [3].

There's a number of data encryption methodologies. The proper methodology should be selected for different scenarios. The below Table 2 listed the data encryption methodologies in different scenarios.

Table 2. Data Encryption Methods and Scenario Examples

Encryption level	Prevention scope	Applicable scenarios
Application level	Malicious direct access to databases	Must be applicable to restricted and confidential data
Database level	Malicious copying of database files (Invalid for authorized database users)	Must be applied in databases containing restricted data and confidential data. It can be superseded by the upper-level encryption
File level	Malicious copying of files (Not suitable for file system administrators and authorized file users)	Must be applied to files containing restricted data and trade confidential data. It can be superseded by the upper-level encryptions
Disk level	Physical loss of disk	Must be applied to the disks out of the physical control of the company and the disks containing restricted data and confidential data. It can be superseded by upper-level encryption

Data Access Control

Data security focuses on protecting data from unauthorized access and corruption throughout the entire data lifecycle. The control of data access is the restriction on the usage of data. Firstly, data can only be used by authorized users, and the unauthorized users cannot use it; Secondly, the authorized users can only use data within the granted scope of permission, and any handling of data beyond the scope cannot be executed.

Data access control can be achieved in the following ways:

Permission Control. Access control is to restrict the users' usage of data. Data can only be used by authorized users, and the unauthorized users cannot use it. In addition, authorized users can only use data within the permitted scope, while not that exceeding the scope.

System Access Control. System access control provides the first level of security protection for the information system. Unauthorized personnel are unable to open the information system through authentication, which means they cannot access data or operate on it.

Data access Frequency Control. For the real users who have already been granted the relevant data access, if they access to the data more frequently than that matches their actual work requires, a rate limiting process should be performed. The common rate limiting algorithms include token bucket algorithm, leaky bucket algorithm, and counter algorithm.

Dealing with External Threats

As a part of network security, data security has always been one of the key targets of network attacks, and the increasing number of external attack methods has brought great challenges and risks to data security and privacy protection. Enterprises should adopt machine learning algorithms to detect anomalies and identify threats as soon as possible with the analysis and warning based on the monitoring of log. The following are some examples of external threats and the measures to address them.

Distributed Denial of Service Attack (DdoS). It refers to the situation that multiple attackers in different locations attack one or more targets simultaneously, or one attacker that controls multiple machines in different locations attacks the target simultaneously using these machines. Due to the fact that the attack targets are distributed in different locations, this type of attack is called a distributed denial of service attack, and there could be multiple attackers [4]. Once a company is caught up in DDoS attacks, the servers will fall into access delay, access failure, or even server unavailability. To reduce the risk, enterprises can consider deploying cloud-based security protection software such as high defense IP.

Phishing Fraud. Phishing attack usually contains malicious attachments. Once being clicked and opened, an enterprise's devices would be attacked, leading to severe incidents of data theft or leakage. So, enterprises should use tools to block phishing emails and use anti-phishing engines. It is also very important to provide anti phishing trainings to employees on a regular basis. In addition, enterprises can also run phishing email exercises to raise the vigilance of the staff. For example, send phishing emails regularly or irregularly to designated or undesignated addresses, find potential weaknesses of security awareness based on the analysis of employees' reading and clicking behaviors, and then deliver the security trainings according to the findings.

Ransomware. Ransomware is a prevalent Trojan virus. It blocks the normal usage of data assets or computing resources with methods such as harassment, intimidation, or even kidnapping user files and extorts money from the users taking this condition. This type of user data assets includes documents, emails, databases, source codes, images, and compressed files, etc. The forms of ransom include real money, Bitcoins, or other virtual currencies [5]. Ransomware is usually combined with data theft. In ransom incidents, hackers may threaten to release the data if ransom is not paid. Ransomware has become the most common security threat targeting various industries and one of the important underground black industries of the Internet. Both enterprises and individuals can become the targets of ransomware attacks and extorting [6].

For enterprises, in order to avoid losses of interests due to ransomware, data security leaders need be proactive to deploy data protection strategies and technologies and to improve monitoring and detection capabilities. Nowadays, most ransomware technology would not be applied for attacks independently. Generally, it would be applied in combination with advanced network attacks, which increases the difficulty of defending against complex attacks. In these circumstances, a multi-layer network security protection system should be built combining a variety of protection methods, such as advanced threat protection, gateway antivirus, intrusion prevention, and other network-based security protection methods.

Network isolation has always been a good technology for network security protection, and it can also be used to defend against ransomware. Besides, it is necessary to backup and restore data on regular basis, since the reliable backup of data can minimize the losses caused by ransomware. But, at the same time, it is also necessary to have security protection for these data backups to avoid infection and damage of data. System updates and security patches, endpoint protection, network segmentation, security software websites, application whitelists, response plan development, and data security backup are all security protection measures. Other measures include terminal protection and monitoring of encrypted network traffic [7].

4 Construction of Data Security System

4.1 Data Security Evaluation

Before developing the data security system, an enterprise should be clear about the data security goals and how to evaluate data security. The security evaluation of data should be conducted from the following aspects: [8].

Compliance. The data is handled on the basis of meeting the requirements of laws and regulations throughout its entire lifecycle.

Reasonable Access Control. Ensure that only authorized personnel can access data.

Data Encryption. Encrypt data according to different data classification.

Data Backup and Recovery. Backup data regularly to ensure that no data loss is caused by accident.

Data Security Audit. Enterprises should make sure that internal audits is conducted on data security and regulation compliance at least once a year. Internal auditors must sit in an independent department to avoid any conflict of interest. Conduct an external audit at least every three years.

Data Classification. Proceed data classification in accordance with the relevant regulations of national information security level protection in a diligent manner.

Employee Training. All new employees should receive data security training within six months of employment. At least two data security related training sessions per year should be taken by all employees.

4.2 Data Security Standards

Laws and regulations provide direction for data security work, while numerous national and industry standards provide specific guidance for the implementation of security work. In order to design data security work and plans for enterprises, data security can be evaluated based on national and industry standards that are suitable for the enterprise.

The enterprise data security department shall conduct comprehensive research on the existing data security work of the enterprise, and identify the corresponding working areas. By conducting preliminary risk assessment and investment in improving security work, the of improvement can be determined.

After setting goals, enterprises can start building a data security system from the following four dimensions: organizational structure, process systems, technical tools, and personnel management (Fig. 3).

Establish a top-down data security management system that covers four levels: decision-making, management, execution, and supervision, and clarify organizational structure and job settings.

Establish a unified data security management process system, clarify the responsibilities of data security work at all levels of departments and relevant positions, and standardize work processes.

Adopting appropriate technical tools to strengthen security management: encryption, desensitization, authorization management, monitoring logs, user entity analysis, data leakage prevention.

Data security awareness education and training. Conduct safety audits on important positions, establish dedicated personnel and separate responsibilities, and if necessary, establish dual personnel and dual positions.

Fig. 3. Four dimensions of data security management

Organizational Structure

Data security governance needs to be carried out in a top-down manner, so it is very important to establish a data security management system that covers decision-making, management, execution, and supervision from top level to bottom level, as well as a clear organizational structure and job settings. Consider appointing a dedicated data security officer in the enterprise, who is fully responsible for the organizational construction of data security. The organizational structure can include the CEO and senior leaders reporting directly to the CEO as the data governance steering committee for the decision-making level. The management team consists of the Chief Information Security Officer, Data Protection Officer, Business Unit Heads, IT Heads, Compliance and Risk Management, and other key management positions. In the execution layer, dedicated team members will take the responsibilities to implement the day-to-day data security work. And the auditors will conduct audit review and supervision independently on specific data security works [9].

Process

At the process level, establish a unified data security management process, clarify the responsibilities of data security work at all levels of departments and relevant positions, and standardize work processes.

Technical Tools

Strengthen security management by balancing risk and investment and adopting appropriate technical tools: to protect data using encryption and data masking technologies, to expend the coverage of existing authorization management systems, to reinforce the standards of log monitoring, to analyze the users and entities based on the logs, to make alerts for abnormal situations and handle them promptly. To prevent potential problems, enhance the usage of data leakage prevention tools, establish prevention and detection mechanisms, and assign dedicated personnel to monitor the alarms triggered by the tools on a timely basis.

Personnel Management

In terms of personnel management, it is necessary to enhance data security awareness education and training. Conduct safety examination on key positions, establish dedicated personnel and separate responsibilities. If necessary, establish dual personnel and dual positions [10].

4.3 Data Security System Construction Plan

Based on the evaluation and analysis and the goals set for data security, enterprises can develop rigorous improvement plans and implement it accordingly. Here is a simple example, and there are a number of data management process systems and documents behind it (Fig. 4).

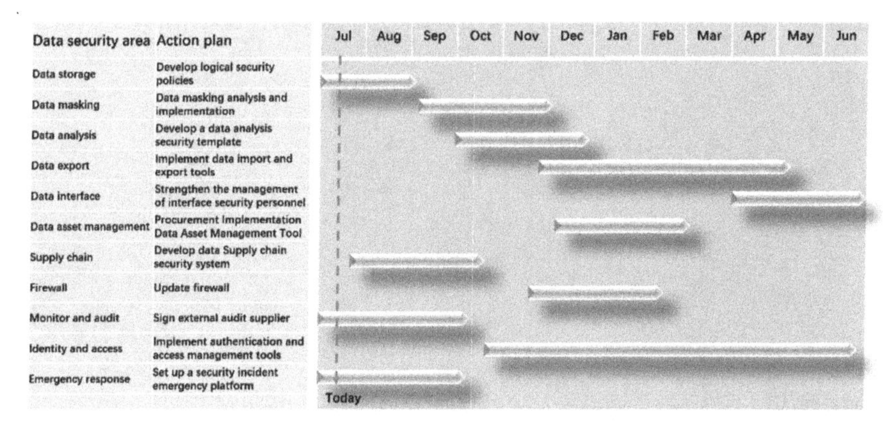

Fig. 4. Example: Developing an improvement plan and implementing it

After the analysis, the data security officer (dedicated personnel) of the enterprise can define the goals of the works following and determine the key points. Then the data security work can be carried out like clockwork.

References

1. Debo, H., Sus, E., Lau, S., Ele, S., Ev, S.: DAMA-DMBOK Data Management Body of Knowledge, 2nd edn. DAMA International, pp. 217–220 (2017)
2. We, L., Liw, Ch., Li, J.F., Yac, Z., et al.: Data Lifecycle Security Specification, pp. 5–7 (2021)
3. Kepu, C., Kexu, B.: Data Encryption. Baidubaike. Kepu China. Scientific Encyclopedia Entry Writing and Application Work Project. https://baike.baidu.com/item/%E6%95%B0%E6%8D%AE%E5%8A%A0%E5%AF%86/11048982?fr=aladdin. Accessed 20 Apr 2022
4. Ling, W.M.: Data service platform: distributed denial of service attacks and preventive measures. In: International Academic conference on Office Automation, p. 1, 20 November 2018
5. Kepu, C., Kexu, B.: Rasomware. Baidubaike. Kepu China. Scientific Encyclopedia Entry Writing and Application Work Project. https://baike.baidu.com/item/%E5%8B%92%E7%B4%A2%E8%BD%AF%E4%BB%B6/5243210?fr=ge_ala. Accessed 06 Jan 2023
6. Baijiahao. https://baijiahao.baidu.com/s?id=1769993208230818350&wfr=spider&for=pc. Accessed 07 Jan 2022
7. Lia, L.J., Bin, W.Y., Du, Q.J., Tian, W.Z,. Fe, M.: Data Security Practice Guide, pp. 155–158 (2022)
8. Ning, W.A., Ka., Y.: Data Security Area Guide, pp. 175–180 (2022)
9. Ladl, J.: Data Governance: How to design, deploy, and sustain an effective data governance program, p. 27 (2021)
10. Li, Z.: Data Governance and Data Security, pp. 105–108 (2019)

Author Index

S. Zhang et al. (Eds.): BigData 2023, LNCS 14203, pp. 231–232, 2023.
https://doi.org/10.1007/978-3-031-44725-9

Printed in the United States
by Baker & Taylor Publisher Services